The Sound of Thirst

*Why urban water for all is essential,
achievable and affordable*

The Sound
of Thirst

Why urban water for all is essential,
achievable and affordable

David Lloyd Owen

Parthian
The Old Surgery
Napier Street
Cardigan
SA43 1ED

www.parthianbooks.com

First published in 2012
© David Lloyd Owen 2012
All Rights Reserved

ISBN 978-1-908069-91-7

Editor: Kathryn Gray
Cover design by www.theundercard.co.uk
Typeset by typesetter.org.uk
Printed and bound by Gomer Press, Llandysul, Wales

The publisher acknowledges the financial support of the
Welsh Books Council.

Contents

Introduction

Leh, July 1991

Did I have an out of body experience? My journal is a bit unclear about this, but the hallucinations were for real. The night before, I had drunk water from a tap at my guesthouse in Leh, Ladakh's capital in the Himalayas. Despite adding sterilising tablets, I could not even hold down a glass of water the next day. Nawang, the proprietor, found a car to take me to the Sonam Dawa Memorial Hospital, and I waited, alternating between apathy and nausea as flashes of green and red encroached upon my vision. Tests were made, and at one point I thought I was looking at myself, slumped next to a vintage Siemens X-ray machine. I do remember asking the doctor what the sheep were doing in the ward. When I came round the next day, a man was writhing in the other bed, crying in pain. A young woman sat at his bedside. She told me her boyfriend's kidneys had failed while they were

1

trekking, probably from dehydration at these high altitudes. As my hallucinations eased I saw there were no sheep, but there were half a dozen sand-coloured rats scouring the floor for breadcrumbs.

Two weeks earlier I had walked out of my job as an equity analyst, covering water and waste management companies for a leading investment bank. I was tired of the constraints of corporate culture and was now setting up an environmental brokerage. Six thousand miles away, I was confronted by a world where the water they were offered threatened people's lives. Outside the city, people who were in the alluvial valleys, and nomads tending their herds across the highland plains knew where to find wholesome water. Their sources were guarded by Klu and Ngiri – the water spirits of the Bon Po who had seeped into this western Tibetan culture and laid down strict rules about keeping the water clean. But there were 10,000 people packed into Leh, a town without water and sewage treatment of any sort. Urban water presented them with challenges a world apart from what their traditions could safeguard.

Twenty years on and progress across the world is thin. Between 1990 and 2008, 1,052 million urban dwellers have gained access to safe water (1,005 million with household connections) while 813 million gained access to improved sanitation. Unfortunately, as the world's urban population grew by 1,089 million over the same period we are in fact going backwards. 2010 was a pivotal year; for the first time in human existence the majority of us live in towns and cities.

One of the chief problems facing water management is that it has become a proxy for other people's conflicts. A decade ago, even the idea of charging for water was a subject of bitter controversy. Indeed, at the 2003 World Water Forum in Kyoto such was the hostility towards reforming and extending water provision that the sole session on financing water was physically broken up by demonstrators. Since then, necessity (and perhaps a sense of shame) has tempered these emotions to the point that reasoned dialogues can begin.

It is astonishing how neglected water and wastewater is when it comes to investment priorities, despite the remarkable returns this can generate in terms of human health, wealth and happiness. I believe that a politician's chief purpose ought to be promoting human happiness, which is probably why I stick to policy rather than politics these days. If this book can encourage people outside the business of water and wastewater to take its management more seriously, then its purpose will have been served.

One of the chief reasons why water policy is not taken seriously enough is the paucity of decent financial data that is freely accessible. For sectors such as power and telecoms, detailed information about networks and how they perform and are financed has been available for decades. For water and wastewater, such information remains an attainable desire. The essence of what is available today will be found here.

When I wrote *Financing Water and Wastewater to 2025: From Necessity to Sustainability* in 2005, gathering reliable data on 59 of our leading nations was an unrelentingly grim process. When writing its successor *Tapping Liquidity: Financing Water and Wastewater to 2029* in 2009, I found the data was not so desperately poor. In the latter book, I looked at the gap between anticipated capital and operating spending in 2010–29 and the funds that tariffs would generate. For the 67 countries I covered, I estimated a funding shortfall of $1.1 to 2.3 trillion depending on the spending scenario. It is essential that this work is carried out lest even greater problems emerge, so the challenge is to encourage both sustainable pricing and funding to close the gap. Indeed, in a recent project for the OECD, I concluded that $7.1 to 8.7 trillion needs to be spent worldwide between 2010 and 2050 in order to develop a universal and sustainable urban water and wastewater infrastructure.

In Wales, the average household pays £411 for their water and sewerage services or 3.0% of their disposable household income. Their water will meet all the relevant health and purity standards and their sewage is fully treated, backed by an infrastructure with a replacement value of £24 billion – reflecting some £6 billion spent on extending, maintaining and upgrading services over the past 20 years. For the developed world, such standards ought to be the norm. Connecting the developing world is just as affordable and possible. The World Health Organization estimates an urban household water connection costs $148–232 per person and household sewerage $193–258 per household, and WaterAid, the water NGO, found basic water and sanitation in Tanzania costs $150 and $50 per person respectively. That is a lot for those living on two dollars a day, but these costs can be defrayed by a variety of means. The benefits they can reap are of another order.

Unfortunately, I doubt if aid to developing countries will ever amount to more than symbolic value when it comes to water and sanitation. This means that countries

4

will need to look towards financial self-sufficiency in order to attain universal and sustainable coverage. It can be done when people are truly engaged in the process and there is an unrelenting drive towards operational and technological innovation and efficiency, allied with eliminating corruption and political interference. These require tariffs and funding mechanisms that combine affordable Full or Sustainable Cost Recovery and the ability to spread the cost of service extension over an appropriate time.

There is a moral imperative here. Amartya Sen's *The Idea of Justice* distils the obligations that power carries to three points. If it is good ('justice enhancing') and it is freely doable ('feasible'), then that person in a position of power needs to take it seriously, since justice depends on capabilities as well as will. Governments and municipalities need to appreciate the moral dimension of serving all of their citizens rather than those with money and influence. For engineers, Richard Bowen's *Engineering Ethics: Outlines of an Aspirational Approach* shows how they can look beyond the technical achievement towards the humane achievement by seeking to make a device operate effectively and preferably to do so commercially. Those who manage the water sector need to be encouraged to embrace innovation, especially where the art of doing more with less can flourish.

You cannot have a civilised and modern society without universal access to water and sanitation. There is no wriggle room here. The private sector has and will play a crucial role in achieving this and I believe that it is morally derelict to oppose the private sector on grounds of ideology. Many of the obstacles facing the water sector stem from ignorance and misinformation: fear and loathing about the private sector, a belief that water should be free, prejudices that water cannot be beneficially recycled and the toxic myth that there is no such thing as climate change. There is a pressing need for people to understand the challenges water management faces in the coming decades and to appreciate that they can be successfully and affordably overcome.

This book is written for anybody concerned about our urban water management malaise today. It is about problems and solving them, so I have avoided discussing academic theory. Even soft water is a hard subject, so there is much data and quite a few tables, but they do have their value. There are also plenty of endnotes, both to substantiate points made and to enable access to the information the book is based on.

Chapter 1

Safe Water and Risk

The word 'water' comes from the Old English *wæter / wæterre / wætre,* which in turn sprang from the Proto-Germanic *watar*; *watar* sprang from the Proto-Indo-European *wodor / wedor / uder-*.[1] Linguists believe that Proto-Indo-Europeans had, in fact, two root words for water: ap- and wed-; the first (preserved in the Sanskrit *apah*) was 'animate' or water as a living force, while the latter referred to it as an inanimate substance. The tension between these two meanings continues to actuate every aspect of water management to this day.

From a swift case of the runs (in fact, a slow, slow walk back to where a loo roll had been carefully stashed in my rucksack) in Hotan, a small town on the southern edge of China's Taklimakan Desert, to a textbook giardia hit outside Karimiabad in north-east Pakistan's Hunza Valley, as well as my lay-off in Leh, I have inadvertently sampled several of the world's waterborne diseases. But that was my choice. I enjoyed going solo with minimal kit in places

7

where you travelled rather than visited, although always with a couple of pouches of cure-alls from painkillers to antibiotics. For those who have to deal with this water day in, day out, anecdotes provide scant amusement. Meanwhile, I was learning about the business of water. Listening to the BBC's World Service in Delhi in 1993, I heard that the British Government, facing a European Court of Justice hearing, had finally admitted that Blackpool's beaches were, after all, bathing waters rather than incidental adjuncts to a holiday resort. In the office, my in-tray heaved with reports about European directives setting standards for urban wastewater treatment and seeking to return rivers to their natural state.

As a child, I was told about my uncle and grandfather's battles with Blackwater fever (a virulent form of malaria where your urine is stained with haemoglobin from ruptured red blood cells) during their respective stints 'East' as a naval medic and a journalist. While long being aware of the debilitating effect such diseases have, I was still dismayed to see how the exoticism of family stories can be remorselessly replayed today on such a global scale. Twenty years ago, I was encouraging companies, banks and investors to look at the opportunities presented by a new wave of environmental concern. There was a decade's worth of business to be done to sustain the world's water services. Many of those challenges are even more urgent today.

All life as we know it has two characteristics in common: it is carbon based (hence the division between organic and inorganic chemistry) and it contains water. We typically consist of 60–65% water by weight for men and 55–60% for women. When we lose 1–2% of our body weight in water (roughly, a litre), we start to get thirsty and suffer, soon becoming confused and tired. By the time we lose 5–6%, nausea, sleepiness and headaches set in, and at 10–15% vision may be impaired, along with muscular function; delirium can occur. After 15% it is usually fatal. Joe Simpson's *Touching the Void* is a gruelling account of how thirst pushes two mountaineers into a series of misjudgements which spiral beyond what most of us could endure into the delirium of extreme dehydration.[2]

Life as we understand it is unfeasible without regular access to water. But water supplies need to be drinkable as well and too often they are not.

A brief history of water and urban living

The more urbane we are, the more water we need – and the more exacting our needs become. Pastoralists and hunter-gatherers live in scattered groups, usually moving with the seasons, thereby minimising the possible contamination of their water sources. Nomads in eastern Ladakh take care not to disturb the Klu-Kkhijil or Nagas (water spirits).[3] This echo of Bon Po animism has been absorbed into Tibetan Buddhism and helps keep their water sources pure and undefiled. A nomad's meat-based diet lowers the amount of water people need when preparing and cooking their food. Pastoralists such as the Ariaal and Maasai in Eastern Africa drink mainly their herds' milk, while fermenting mares' milk in Mongolia makes Airag (also known as Kumis in Kazakhstan and Kyrgyzstan) a mildly alcoholic drink central to hospitality when living on the move.[4] This lightly watered way of life works in a world far more sparsely populated than ours. Ladakh's Rupshu covers 5,500 square miles and is lived in by three nomadic tribes with a year-round population of 1,200.[5] That might sound scarcely believable, until you realise that this was the approximate population of the British Isles in 9,000 BC.

At that time, the first formal settlements were emerging, such as Tell Qaramel in Syria, Jericho in the West Bank and Göbekli in Turkey, with Turkey's Çatalhöyük housing between 5,000 to 10,000 people from 7500 to 5700 BC. Urban living as we understand it started in 3500 BC with the Sumerian cities of Uruk and Ur – cities that combined administration, trade and manufacture, creating livelihoods for tens of thousands. The Harappan civilisation of the Indus Valley in Pakistan enjoyed universal household sewerage (conduits taking wastewater from households to a distant location) and household loos in Mohenjo Daro and Harappa from 2600 BC. Treatments to improve the taste, smell and appearance of drinking water extend back to the earliest written records, with sunlight (ultraviolet light treatment today), charcoal filters (granular activated carbon), boiling (multiple flash distillation), straining (filtration media) and coagulants (flocculation) mentioned in ancient Egyptian, Greek and Sanskrit texts between 2000 and 1450 BC. While they responded to pressures on their drinking water resources, the people living in these areas today are only now starting to reconnect with their noble pasts.

So people have been living in urban societies for over five thousand years and for the most of that time have been treating water with charcoal and sand filters which are still used today. They have also been connecting houses to urban sanitation systems. Yet household sewerage is very much the exception in the Upper Indus Valley today. Indeed, 72% of urban dwellers in Pakistan in 2008 have some form of household sanitation, and that figure has slipped from 73% in 1990, while the percentage actually enjoying household sewerage is appreciably lower; a survey in 1998–99 found that just 51% of households in Pakistan's major cities had underground sewerage. It is a lot lower today in Gilgit, the region's main town. Globally, official surveys suggest that 21% of people living in urban areas didn't have household water in 2008 and had to rely on untreated water; 24% lacked household sanitation. As a result, 2.4 million people die from waterborne disease every year.[6] Meanwhile, the developed economies of Europe, North America and East Asia face all manner of problems when it comes to sustaining their water resources.

This is what this book is about: trying to understand why all of our urban water and sanitation services are in such a poor state, how our water resources face all manner of new challenges from demographic to climate change and how we can make our urban water services economically and environmentally sustainable. To do this, we need to appreciate the peculiar characteristics of urban water management, why water and sanitation have slipped so far down our collective spending and political priorities, and how something as universal as water is so localised when it comes to challenges and responses.

After 5,500 years of urban living we are a long way away from connecting our cities to sustainable water supplies, let alone sanitation. In Northern Pakistan, for example, it is astonishing to find that sanitation services were far better 4,600 years ago than they are today.

What we need water for

Broadly speaking, managing urban water revolves round municipalities, commerce and industry; rural water is concerned with agriculture and those who live from the land. The need for urban water and its uses break down between consumption (drinking water and the processing, preparing and cooking of food), hygiene (preventing disease through personal and domestic cleanliness), amenity (whether washing cars, filling swimming pools or watering golf courses) and economic (water for industry and commerce from mining to steelmaking to preparing microcircuits).

Consumption varies with circumstances, so people living in hot climates[7] and doing manual work need to drink a lot more water than office workers in cooler climates. A barest minimum for human survival should be two to three litres a day for adults (some of which already comes from food) and two litres per day for basic food preparation. In developing countries, at least 7.5 litres ought to be needed every day, but in East Africa one survey found that daily use was just 4.2 litres where people have piped water and 3.8 litres where they do not have a household tap.[8]

Policymakers split water use into three categories; agriculture, industry and municipal.

Agriculture accounted for 66% of all the water we extracted in 2000, having fallen from 89% in 1900.[9]

Here the challenge lies in reconciling availability and reliability of supplies with an ever-increasing population and their new expectations, especially as more water-intensive 'western' diets have been adopted. Then there is the matter of maintaining soil quality when addressing issues such as salinity and topsoil run-off. From the Nile to the Indus, yields in once fertile soils are declining as salt levels build up because river modification prevents occasional flooding to wash excess salts away.[10] In China, 4.5 billion tonnes of topsoil are lost every year due to erosion, with dust storms reaching as far as the USA. To give you a flavour of the ever-evolving difficulties facing agriculture, 13 billion hectares of the earth's land outside the Arctic and Antarctic Circles were surveyed in 1991. Of these, 1,660 million hectares were classified as lightly to moderately degraded, while 305 million hectares had suffered strong or extreme degradation by salination, topsoil loss or pollution, meaning they are effectively beyond practical reclamation.[11] It is easy to be pessimistic, but agricultural challenges can and will be met; between 1990 and 2004 agricultural water efficiency improved by 1% per annum. This commendable progress has been blunted, however, by population growth, but there is great scope for further improvements and mobilising neglected resources such as wastewater.[12]

Industry accounted for 10% of water withdrawals in 2000, up from just 4% in 1900. Managing industrial water involves a number of disparate drivers, ranging from the need to be more efficient as water is increasingly seen as an economic good and as companies' water supply and industrial effluent compliance costs mount. Certain sectors also seek dedicated suppliers of ultrapure water, either through a treatment unit provided and serviced by a specialist or managed as part of an outsourcing package. At the same time, as water abstraction and effluent discharge regulations and their regulators start to bite, so do campaign groups, looking for targets. Industrial customers look for reliable supplies, service efficiency and value for money. This can also be seen as a conflict between people's right to access to water to survive and a company's perceived right to do business.

Municipal and domestic water use accounted for 20% of water withdrawals in 2000 and a further 4% for reservoir storage, against 8% for both in 1900. While expectations can vary hugely (people can seemingly take poor service for granted if that is all there is on offer) there are general expectations about reliability (water flowing out of taps when you want it), quantity (getting enough water), potability (water that is safe to drink) and quality (how it tastes, smells and looks). As humanity gathers in towns and cities, demand for water becomes more concentrated, often outstripping local rivers, calling for increasingly complex supply management. This means that major cities may rely on water piped in from beyond their river basins, and as water is withdrawn from an area appreciably larger than the city, the city's 'water footprint' grows.

Rising expectations about water quality and quantity increases the disconnect between cities and what their watersheds can provide and indeed, much of the water consumed may be 'embedded' in the food a city imports and in the goods a city imports.

Municipal and industrial water demand is propelled by broad yet diverging narratives. Urban water consumption has been easing for the past 30 years because of improved efficiency among developed economies countering the effect of new consumer goods such as dishwashers and power showers, while in developing countries, a growing population with poorer water access drives down overall average use. In contrast, water use by industry is set to rise as the demand for consumer goods and manufacturing output for urban and rural communities has outpaced improvements in industrial water efficiency.

From Bangladesh to Baltimore there is more in common about long-term needs than many may suspect. We cannot discriminate on access and people have a right to reasonable expectations (100 l/cap/day before recycling and reuse), but when developing these systems where none previously existed, we can engineer in our experience through a plethora of lessons we have learned elsewhere.

I am concentrating on urban water and sanitation because I believe that the various issues facing agricultural water and rural water and sanitation are appreciably better understood than those affecting urban areas. We have, for the first time in our existence, become a predominantly urban species and so the timing is appropriate.

Access goes beyond slaking our thirst

In 2000, the United Nations unveiled its eight Millennium Development Goals (MDGs), designed as targets for improving sustainability and human dignity over the next fifteen years. MDG Number Seven included the pledge to 'halve, by 2015, the proportion of people without sustainable access to safe drinking water and sanitation' against a 1990 baseline. In fact, water and sanitation lie at the heart of seven of these goals. Much extreme poverty and hunger (MDG 1) is water related, while poor access to water means children have to collect water rather than going to school and that others miss school due to illness (MDG 2); women have less opportunities to go to school due to water-gathering duties (MDG 3); child mortality is often the result of water related malnutrition and illness (MDG 4); a lack of clean water impairs maternal health, notably during childbirth (MDG 5); and waterborne illness leaves people more vulnerable to other illnesses and can redouble their harm (MDG 6). Putting these into perspective in MDG 2, one of the chief reasons for 121 million children not even attending primary school is that they are needed at home to fetch water and 443 million school days are lost every year due to waterborne diseases.[13]

It is clear that the significance of water extends far beyond simply quenching our thirst, to health and education. Access to safe water is about having the hygiene necessary for human health, comfort and dignity, from hand washing to cleaning your hearth and home. Likewise, sanitation is about keeping sources of infection away from where we eat, live and sleep. The World Health Organization has a sliding scale of definitions for water access.[14]

Those with no access often collect less than five litres each every day, having to walk more than a kilometre to fetch their water, a return journey lasting at least half an hour every day. They cannot be sure of getting enough water to drink, let alone whether it is fit to drink and unless they wash where they collect their water, safe hygiene isn't possible. Their health risk is very high.

People with basic access have to get by with less than 20 litres each day, spending five to 30 minutes collecting water and getting at least enough to drink. Hand washing and basic food hygiene should be possible, but laundry and bathing are not feasible unless carried out at the source. Their health risk is high. Interpretations vary widely; India's definition of basic access is one public tap per 30 households (which works out at 162 people) at 40 litres a day each.[15]

At the intermediate access level, people get an average of about 50 litres per day, with water delivered through one tap on their plot, or less than 100 metres away. Consumption, along with all basic personal and food hygiene is assured and laundry and bathing should also be assured, and the health risk is low.

Optimal access means getting at least 100 litres each day, via household taps, which ought to be working continuously. All consumption and hygiene needs should be met and health risks are very low.

Between 1990 and 2008, urban household access to drinking water rose from 41% to 52% in South East Asia and from 87% to 96% in East Asia, but it actually fell from 43% to 35% in Sub-Saharan Africa and from 55% to 51% in Southern Asia. Access has a huge bearing on how much water people can enjoy; researchers at Loughborough University's WEDC found that in Jinja, Uganda, water usage is closely related to access. Those with communal springs, handpumps or standposts used 16 litres a day, against 50 for an in-compound yard tap and 155 for household taps.[16]

In 2000, 18% of people worldwide had no access to safe water, 30% to basic improved sources less than a kilometre away and 52% through household connections.[17] At anything less than intermediate access, people's quality of life and the possibility of a better life are being held back by their lack of water. This means that half of humanity are having their lives stunted by their lack of access to water.

Common sense and robust definitions about suitable levels of access for water are needed in order to concentrate people's minds on delivering real benefits that high quality water access can offer. It is not an issue of national pride, but an issue about people's dignity and common decency.

Sanitation is another story

Nulla vita sine aqua; sine cloacis vita brevis

Without water there is no life; without drains life is short. This is a motto Colin Drummond drummed up when Master of the Water Conservators, one of the City of London's Livery Companies. At our functions, the chief toast is taken in water, and our badge is based on a drop of water. Money to support water education and research is collected in a replica water carrier.

When it comes to human health, bacteria are both a blessing and a curse. The bacteria that abound in our colons perform a crucial role in turning digested food into sewage and breaking this down into its essential components, feeding the nutrient and energy cycles that sustain life on earth. Unfortunately, we have a very limited tolerance to them in the rest of our digestive systems. In addition, the intestinal parasites ranging from protozoans to tapeworms that live inside us for at least a part of their lives are spread by excreta when it is not properly treated and disposed of.

The greatest difficulty facing sewage treatment and sanitation is one of perception. The connection between access to safe water and well-being is simple and straightforward, but without education, the case for sanitation spending is more convoluted. Sanitation is about the 'five Fs'; faeces, fingers, flies, fluids and fields. Until these enter the popular imagination, sanitation is set to remain the water sector's 'poor cousin' which, when you consider the state the water sector is in, is not good news. Does it matter? Perhaps so, when you realise that improved sanitation was the chief driver behind a fifteen-year rise in Britain's life expectancy between 1880 and 1920, or in the words of the *British Medical Journal* it has been the 'most important medical advance in the past 150 years.'[18] In Sub-Saharan Africa, the 2015 sanitation MDGs are on target to be met... in 2076.

But merely meeting numbers does not deal with the job at hand. While everybody who gets connected to a household tap directly benefits from this, as far as sanitation is concerned the real benefits only start to kick in with near-universal improved access. If 10% of people in a slum are left with inadequate sanitation, their excreta is quite enough to befoul the lives of the connected households. This is hardly helped by the target mentality when well-meant exercises such as the Millennium Development Goals become ends in themselves, rather than an opportunity to consider what they actually achieve. For example, WaterAid is concerned about 'latrinisation', the installation of household latrines simply to satisfy development goals, rather than how they are meant to deliver safer sanitation.[19]

Another challenge lies in the linkage between access to water and sanitation as they are separate, yet intertwined. The more water a household has access to, the greater its sanitation options.[20] There is little point in flush lavatories if there is not enough water to flush the waste safely away. Households 'served' by a communal standpipe get by with 15–30 litres per person per day, compared with 25–70 for yard taps and intermittent household connections and 90–180 and upwards for in-house connections with a full supply. So, where households have 'improved' sanitation but limited water, further progress depends on improving water supplies as well. Technology pushes sanitation the other way as well, with modern dual-flush lavatories using 20–40 litres of water per person each day compared with traditional full flush models, which put away 30–60 or more litres every day. All of this is splendid – as long as the sewerage network can work with these lesser flushes. Sometimes, they cannot and sewer blockages are on the rise.

While household water connection rates in major cities were between 96–100% for Europe and North America, they fell to 73–77% in Latin America, Asia and Oceania and 43% for Africa. The differences for household sewerage were starker still at 92–96% for Europe and North America, 35–45% for Latin America and Asia and 15–18% for Africa and Oceania.

Although sanitation is just as important as water, it

does not get the same attention. Yet poor sanitation is one of the chief causes of poor water quality. Perhaps one of the problems is that 'sanitation' is a rather genteel word. If sewage or excreta were used instead when writing about access, it might concentrate minds more.

Who lacks safe water and sanitation and why

Since 2000, the Millennium Development Goals have made some headway towards reducing the number of people without access to improved water and sanitation, but such have been the effects of population growth and urbanisation, that it has often been a case of running hard to stand still. Between 1990 and 2008, the number of people living in urban areas rose by 110 million to 772 million in the developed world and by 969 million to 2,396 million in the developing world. In the developed world, there appears to be little to worry about, with all (in theory) enjoying improved sanitation and 98% having household water connections throughout the entire period.

In the developing world, those with access to improved sanitation rose by 701 million to 1,629 million (68%), those with access to improved water by 925 million to 2,252 million (94%) and those with household water by 736 million to 1,749 million (73%). This ought to sound pretty encouraging until you put it another way; the number without improved sanitation in fact rose by 268 million, those without improved water sources by 44 million and those without a household tap by 233 million. It means that while globally we remain on target to meet the water goals by 2015 (although not perhaps in Sub-Saharan Africa and South Asia) for sanitation, the currently projected outcomes from the 2008 data point to 2.7 billion unserved in 2015, against a target of 1.7 billion by that time.[21] There is little time left to close such a gap, and so already, concern is turning to what needs to be done after 2015.

Access to available water is what matters. The OECD uses four tiers of access to water. For urban areas, it believes 135 million people collect unsafe water, 580 million more collect safe water, 1,770 million have piped water, with a basic level of regulatory oversight, and finally there are 750 million who enjoy piped water provided by fully regulated utilities. This means that the lives of 715 million urban dwellers could be significantly improved through water connections. The actual quality of the water provided appears to have been overlooked. Gerard Payen has taken a cool look at what the 2008 connection data may mean on the ground. He concluded that while 5.8 billion out of the world's 6.7 billion people may have had access to 'improved' water sources in 2008, perhaps 3.3 billion in fact had access to safe water, 1.9 billion with water of doubtful quality and 1.9 billion depended on water of hazardous quality.[22]

With over 95% of people living in high-income countries having satisfactory access to potable water and appropriate sanitation services by 1990, along with 74% access to potable water and 68% access to appropriate sanitation by 1990 in medium-income countries, it is little surprise to find that public and political concerns in the more developed economies are increasingly being driven by environmental and aesthetic considerations, while those in the less developed economies remain rooted to those of the most basic public health.

Access is strongly related to wealth and place

Anybody who believes that the current service and access paradigms are equitable needs a brisk reality check. Surveys carried out in 2000 to 2005 found that in Namibia, 98% of the richest 10% have flush lavatories while 100% of the poorest have no safe access to lavatories whatsoever. [23] In Colombia, all of the wealthiest 10% enjoy flush lavatories and while 41% of the poorest also have them, a further 41% have no safe sanitation. While all of Peru's top 10% have piped water, just 24% of the poorest do and 65% depend on water vendors and 91% of the wealthiest in Benin have piped water against 13% of the poorest.

At many a water-policy gathering, there will be a solemn reiteration regarding access to water as a human right. As with gender equality, freedom from torture and freedom of expression, such sentiments can attract hollow laughter beyond the gilded halls, tinged darker by the way sanitation slips their minds. Sometimes the desire is there; after all, should ensuring the essentials of human decency really threaten the elites? Often it is a case of priorities. Water is not a 'sexy' subject, let alone its 'poor cousin'. But above all, it is a failure to transmit national and international obligations and desires to cities and municipalities in a compelling manner.

The Millennium Development Goals were meant to be a start. Yet we are in some danger of not even reaching the start, as the pace and scale of urbanisation overwhelms many connection initiatives. So the need to go from this to what really matters, *universal access*, gets more urgent as time passes.

Mega cities and mega slums

In 1988, I took the train for my first trip from Hong Kong to Guangdong. Two great cities separated by the New Economic Zone of Shenzhen, banner-blazoned communes, small towns and lush vegetation. Twenty years later, I sat inside the evening ferry and both sides of the Pearl River were bounded by ribbons of lights snaking along and up every slope. The two cities had melded into an agglomeration of 120 million people.

What is an urban area? It depends on the country. It ranges from 200 people in Iceland to 10,000 in Senegal, with the rule of thumb being, for example, 1,500 in the UK and 2,500 in the USA all living in a distinctly settled area. The European Union, when drawing up the Urban Wastewater Treatment Directive drew the line at 2,000 people. It is easy to underestimate the sheer diversity of urban settlements, especially in developing economies. Every locality has its own circumstances, ranging from tenure and population density to what (if any) infrastructure is already there. A town or village may be subsumed into urban sprawl, presenting a range of housing densities, street layouts, open spaces and incomes. Multi-storey housing needs flush sanitation, while there are always issues about land ownership and poorer communities, especially when it comes to financing sustainable infrastructure.[24]

We live in the age of urbanisation in general – and the mega city in particular. In 1950, two of the world's 30 largest cities had more than ten million people and 19 of them were in the OECD (the Organisation for Economic Co-operation and Development), a grouping of 34 advanced economies. By 1990, ten of the largest cities had more than ten million people and nine were in the OECD, with 21 having more than ten million in 2010 and nine were in the OECD. The UN forecasts that 29 will house more than ten million in 2050 and seven will be in the OECD. While the 30 biggest cities had 118 million inhabitants in 1950, this had grown to 406 million by 2010 and is forecast to grow to 479 million by 2025. London was the world's largest city in the Victorian age and even in 1950 it was the third largest. Indeed, four of the world's largest cities were in the UK back then. By 1965, London was the UK's sole representative, slipping to 19th place by 1990 and 30th place by 2010. It does not feature in the forecasts.

The challenge here is that the further the mega cities shift away from developed economies, the further they slip down the developed world's political agenda. That cannot be a good thing when it comes to infrastructure spending. Water and sanitation infrastructure is one of the chief constraints facing the development of mega cities. None of the cities forecast to become mega cities in the next few decades have adequate water or sewerage services.

The scale and pace of urbanisation is repeated across the board. There were 74 cities with a million people in 1950, 265 in 1990, 441 by 2010 and a forecast 546 by 2050. Urban living has swung from being the exception to the norm, from 29% in 1950 (53% in developed countries against just 18% in developing economies) to 50% by 2010 (75% and 45% respectively) to 69% forecast for 2050 (86% for developed and 66% for developing economies).

Population estimates and projections drawn up by the United Nations give an idea of the sheer scale and trajectory of urbanisation. Africa and Asia stand out for the sheer velocity and scale of their urbanisation.[25] Over a century, the urban population of Africa is forecast to grow 37 fold, while there will be 3.15 billion more urban dwellers in Asia. Africa's urban population grew by 370 million in 59 years from 1950. It is projected to grow by a further 832 million in 41 years from 2009, a tripling of the rate. Urbanisation at such a rate and on such a scale does rather put the current angst about South East England future population into a broader context.

A century of change – the human shift to urban living, 1950–2050

Urban population (million)	1950	2009	2050
Africa	33	399	1,231
Asia	229	1,719	3,382
Europe	281	531	582
Latin America	69	462	648
North America	110	285	404
Oceania	8	25	38

What we understand as an urbanising world is also changing. Asia and Africa's share of the world's urban population shifts from 35% to 74% while Europe and North America's falls from 47% to 15%. In 100 years, a greater population shift is happening than over the past 5,000 years of urban history.

Africa and Asia take over as the main urban areas

% of the world's urban population	1950	2009	2050
Africa	4%	12%	20%
Asia	31%	50%	54%
Europe	38%	16%	9%
Latin America	9%	13%	10%
North America	15%	8%	6%
Oceania	1%	1%	1%

While the mega cities grab the headlines, other patterns are just as interesting. The biggest segment is the towns and villages growth is almost stagnant at the next step up. This is mainly due to smaller cities growing and being promoted into the larger segments, along with people moving first to local cities and then to larger ones as they become steadily more immersed in urban life.

Who lives where by city population

Urban population by size (million people)	1975	2009	2025
10 million or more	186	320	469
5 million to 10 million	143	225	321
1 million to 5 million	546	749	1,004
500,000 to 1 million	237	352	465
100,000 to 500,000	534	629	684
Less than 100,000	914	1,146	1,593

Strangely nobody has made a serious estimate about the running total of urban areas worldwide, but if we draw the line at 5,000 people, then Columbia's GRUMP, the Global Rural Urban Mapping Project identified 24,000 such urban areas in the year 2000.

We are experiencing the most dramatic process of urbanisation in our history. This is taking place across the world and across all types of towns and cities. It is also happening at its greatest rate and scale where people are the most poorly placed to meet the challenges urbanisation brings.

A slum future?

The United Nations Human Settlements Programme, known as UN-Habitat, has been studying the world's towns and cities since 1978. One of its achievements has been to focus on the greatest of urban taboos: slums. Typical synonyms for 'urban' are elegant, refined and cultured, which have little to do with slum realities. I have little time for 'euphemasia' (people dying of neglect when an awkward subject gets avoided), which is why slums are slums, not 'informal settlements', just as developing economies are not 'countries of the South'; when current realities are unacceptable, they need to be recognised as such. Language ought to reflect reality – the best way to get rid of the word 'slum' is to turn it into an anachronism. A recent UN report noted that in 2001, 926 million people, or 32% of the world's urban population lived in slum areas and while 43% of the urban population of less developed economies live in slum areas, just 6% do so in developed economies. The UN anticipates this figure rising to 2 billion by 2033 and 3.5 billion by 2050.[26] In other words, slums are growing by 70 million people a year as a result or rural to urban migration, with the proportion of humanity living in slums rising from one in six today to more than one in three in less than 40 years time. This is very much a regional concern, for despite the media publicity about Brazil's favelas, 72% of the urban population of Sub-Saharan Africa live in slums along with 58% in South & Central Asia, compared with 24% to 36% in other developing regions. These simple figures retain the potential to knock the UN's water and sanitation MDG targets into the dirt.

From slums past to slums present

One of the problems we face is that of attitude. The United Nations defines a slum as an urban population living without at least one of the following: access to improved water, access to improved sanitation, sufficient living area, durability of housing and security of tenure.[27] By that definition, slum dwellers are condemned to inadequate water and sanitation services. UN Habitat estimated that there were 737 million people living in slums in 2005[28] (the numbers do tend to bounce about a bit) yet as we have seen more urban dwellers were assessed as lacking adequate water or sanitation. To give some extreme examples, I have used data for household access to piped water and sanitation and compared these with slums:[29]

Differing realities – official estimates of the urban population in slums (2005) and having household water or sewerage (2008)

	All with household		Live in
	Water	Sewerage	Slums
Cambodia	57%	46%	72%
South Africa	89%	78%	27%
Tajikistan	83%	45%	94%

What can these numbers tell us? In theory, you would expect few slum dwellers in, for example, Tajikistan to enjoy household water and sewerage and therefore the proportion of the urban population with access to both of these to be little above 6%. That evidently is not the case.

The way these numbers grind into each other is worrying, as they imply that we may have interest groups reporting (or indeed developing) numbers for their own purposes which are in fact obscuring what can and needs to be done. Agency ought to talk to agency. As we will see in later chapters, schemes in Manila, Jakarta, Phnom Penh and Morocco show their desire for household water connections and their willingness to pay when given a chance. People upgrade their houses when conditions allow, moving from cladding to bricks and mortar. Privacy and dignity matter and pipes and sewers are part of these aspirations. Challenges such as genuinely universal access to water and sanitation are best met when knowing what your targets are meant to be. These numbers demonstrate that unless proven otherwise, slums can and ought to be connected to those services which we take for granted.

All problems facing urban water and sanitation are at their most intense in these slums. These problems can and must be confronted. As Robert Neuwirth has pointed out, slums are not going to go away and, indeed, many of the West's leading cities from London to San Francisco had large slum areas a century ago. What is needed is to bring them into the formal economy, through security of tenure (you might not own the land, but at least you have protection from arbitrary eviction) and avenues of participation into the city's civic society. Such conditions encourage slum dwellers to invest in basic services such as water and sanitation.[30]

Water and health

Lacking access to safe water and sanitation can be debilitating or deadly, especially in developing economies. The World Health Organization has a four-step classification of water-related diseases:[31]

Waterborne: From faecal contaminated water (e.g. dysentery, diarrhoea, cholera and typhoid)

Water-washed: Spread due to insufficient washing water (e.g. trachoma and scabies)

Water-based: Organism spends part of its life in water (e.g. schistosomiasis)

Water-related: Insect vector spends part of its life in water (e.g. malaria)

The first two are directly related to access to drinking water and the safe disposal of sewage. The latter two are caused by people living near to unhealthy stretches of water or, more to the point, water which is shared by creatures that harm human health.

The way we trade affects the way diseases a spread. Pathogenic or disease-causing microorganisms have long been recognised as the cause of waterborne diseases and, as we will find out in chapter four, how they spread within cities. But how do they get there in the first place? In 1849 John Snow wrote that 'epidemics of cholera follow major routes of commerce. The disease always appears first at the seaports when extending into islands or continents', which remains the case today, although there are also new trades, ranging from refugees crossing land

and sea to jet-setting tourists that are increasing the speed and intensity by which diseases such as SARs and Bird 'flu are spread.[32] This is one of the reasons that between 1970 and 2000, the World Health Organization found that 35 new waterborne pathogens have been recognised or found to be re-emerging, ranging from new varieties of Hepatitis to Ebola Fever and Legionnaire's Disease.[33]

Death attracts headlines. Chronic and acute illness can hurt even more and beyond death lays a hierarchy of loss, from time lost due to a minor ailment to debilitating illness that renders people economically inactive. A DALY, the acronym for 'disability-adjusted life year' refers to the equivalent of one person being unable to work for a year due to illness. This, in the language of economists, is 'opportunity cost', the opportunities lost by a lack of access to water and sanitation.

John Maynard Keynes was right to remind us 'that in the end we all die', but getting to that end in a state of some dignity remains something of a western luxury. Savings are a cushion against misfortune, but the worse off you are, the less money there is for a 'rainy day' and the more chance of you having one. For people living beyond society's safety nets, illness may expose you to debt and to money lenders. You may lose your job, your home or your family. All of these factors can propel people from being merely poor to facing absolute poverty.[34]

Every year, 1.4 million children die from diarrhoea along with 100,000 adults. A further 860,000 die from malnutrition brought about via waterborne disease and 31,000 through other waterborne diseases. On top of this 2.39 million, a further half a million die from malaria, which is a water-related disease. Two billion people worldwide suffer from intestinal nematodes such as hookworm, 200 million from schistosomiasis flatworms, 25 million from lymphatic filariasis and five million from trachoma. In total, at least 9.1% of all DALYs and 6.3% of all deaths worldwide in 2002 stemmed from inadequate water, sanitation and hygiene.[35] One study estimates that diarrhoea through poor water and sanitation accounts for 54 million DALYs or nearly 4% of all diseases.[34] Assuming that the average person affected earns $2 per day (the UN's definition of the poverty line), each DALY costs $700, which works out as $38 billion lost through illness alone even before we consider the cost to their families.

If you live in Europe and North America, the effects of poor water quality are closer than you might believe. In May 2000 the public water supply of Walkerton, a town of 5,000 people in Ontario, Canada, was contaminated by various pathogens when animal waste leached into a well. Between 2,300 and 2,500 people fell ill and seven died as a result. Research afterwards revealed there had been 99 disease outbreaks in Canadian public water systems between 1974 and 2001 and many more in semi-public systems. Meanwhile an outbreak of cryptosporidiosis in Milwaukee, USA in 1993 saw 400,000 people falling ill and 50–100 deaths, the latter principally among AIDS patients whose immune systems were compromised, at a total cost of $96 million. The outbreak was not detected by the surveillance systems until it was too late. This threat is a new one as cryptosporidium was only identified as a human pathogen in 1976. Even so, it is already widespread, with 20% of young adults showing evidence of infection in the USA and more than 90% of babies in a Brazilian shanty town.[37] The next really serious 'crypto crisis' is only one system failure away.

Just as water and sanitation are the neglected elements of urban infrastructure, waterborne diseases tend to be neglected. While the 'big three' diseases AIDS, tuberculosis and malaria grab the media headlines and research funding, diarrhoea kills more than they do combined, 90% from children under the age of five and 73% in just fifteen countries. Dealing with waterborne disease has been a recent achievement in the western world; 40% of schoolchildren in the southern states of the USA had hookworm in 1900 and its elimination resulted in improved literacy and higher incomes. Connecting hand washing to water and sanitation is crucial, since poor access to water reduces people's ability to wash their hands regularly, while improved sanitation minimises the spread of diseases in the first place.[38]

Waterborne diseases are not glamorous, but they make up for this in their ability to despoil lives and livelihoods. New diseases are also being identified, broadening the need for appropriate urban water management.

Water and the collapse of cities

Water makes and changes urban history. In recent years there has been a vigorous debate about lead in water causing the decline and fall of the Roman Empire.[39] Lead poisoning from food preparation, wine amphorae, lead water pipes and storage tanks may not have brought about the decline and fall of the Roman Empire, but lead levels found in skeletons from patricians to slaves were certainly high enough for it to have played some role.[40] Today, concern is focussed on the effects of lead in children's mental development, which is why lead has been progressively phased out of petrol and paints in recent decades.[41] In plumbsolvent areas (those with soft water) even lead solder in otherwise lead free pipes can be enough to make the water fail the new European standards, which have been made 30 times tighter between the 1970s and 2013 when it is estimated that 120 million people living within the European Union will fall foul of these new standards.[42] Perhaps the Roman debate has been misplaced, and instead of considering poisoned adults the real problem was of cities being reduced to moronic infernos by generations mentally impaired through rising lead levels.

In his book *Collapse,* Jared Diamond notes that problems with water management are one of the eight key drivers causing once mighty civilisations and societies to implode.[43] There have been plenty of cases where the loss of water supplies could not be endured. No tongues wag in Babel, nor is Sheba fit for a Queen, bearing gifts or otherwise. Great cities pass into history without water to sustain them.

Cambodia's Angkor Wat was one of the largest pre-modern cities, flourishing between the ninth and fifteenth centuries. Annual dry seasons and monsoons meant that the city's expansion was dependent on water channels and reservoirs, but when these became unviable due to erosion and sedimentation (perhaps through complacency or neglect), the city was unable to endure a prolonged drought and collapsed. Likewise, water scarcity including three severe droughts is amongst the drivers for the collapse of the Classic Maya cities in 810–910[44] and a number of urban societies in the Middle East, although Harvey Weiss and Raymond Bradley do caution that particular reasons fall in and out of fashion, and it is clear that climate change is currently a fashionable approach.[45]

Can it happen again? That is up to us. Mexico City and Beijing are regularly touted as the next cities to collapse. In Mexico City's case, water scarcity is combined with lowering groundwater causing the city to sink by up to nine metres in the past century, further damaging the city's literally creaking water and sewerage infrastructure. Beijing suffers from over-abstraction of both surface and groundwaters, with groundwater levels falling by 12 metres between 1960 and 2000 and people having access to one eighth of China's average water resources, and less than a third of what is defined as the level of severe water shortage.[46]

Yet Mexico City and Beijing carry on. Doomwatch prophesiers tend to depend too much on straight line projections at the expense of appreciating how cities adapt as circumstances change. When politicians decide to shift to shiny new capitals, whether from Rio to Brasilia in Brazil or from Alma Ata to Astana in Kazakhstan, the old cities don't just fade away. They were there for a reason and have a habit of carrying on, albeit with fewer diplomats, politicians and their allied professions.

There is also the issue of scarcity in regions and countries as well as cities. We will take a good look at this in chapter three but in the meantime, perhaps we should consider many cities as isles of water scarcity in seas of relative plenty.

Cities and civilisations do not last forever. In quite a few historic cases, their collapse can be linked to changes in water supplies or its management even when water was available.

Threats are expensive and extensive

Measuring how humanity is threatening our water resources remains a work in progress. Just as each city's hydrology is unique, so is the way each city's activities impacts its water resources. Since it was first mooted in 1991, the European Union has been inching towards a Water Framework Directive whereby the inland waterways of the EU member states will be of 'good ecological status' by 2015 (in fact anything achieved by 2027 will be very welcome). This has resulted in unprecedented work to find out exactly what condition rivers are in, rather than making assumptions based on some spot observations. It remains an incomplete art; one consultant, Colin Fenn, told me about a conference in 2007 where presentations about river water compliance in France and Belgium made it clear that water quality improved by a factor of three merely by flowing across one border.

The marriage of remote sensing, supercomputers and analytical modelling is transforming the way we look at the human and natural environment and how they are changing. In recent years, many of the breakthroughs have been in moving from broadly based analysis to examining events in relatively small areas. Charles Vörösmarty, a specialist on global water resources, and Peter McIntyre, freshwater zoologist, looked at 23 threats to water resources covering watershed disturbance, pollution, water resources and biotic factors by dividing the earth's landmass with rivers into 46,517 data cells.[47] They estimate that 4.8 billion people across the world faced a 'very severe' threat of water scarcity or water biodiversity loss in 2000. These threat levels were then adjusted to take into account each area's water and sewerage infrastructure, which lowered the number of people under threat to 3.4 billion. This turned Western Europe and the USA from the most threatened to amongst the least threatened, as they benefitted from 'massive investments in water infrastructure, the total value of which is in the trillions of US dollars.' This is starkly illustrated by Vörösmarty and McIntyre's model which found that 43% of people in low-income countries (GDP less than $1,000 per person) faced a human water security threat, but once this was adjusted to take account of their water and sanitation infrastructure, this rose to 96%. In complete contrast, while 90% of people living in high-income countries (GDP over $10,000 per person) faced human water security threats, this fell to just 5% when infrastructure was considered. For example, Ethiopia has 150 times less reservoir capacity per capita than the USA, despite its being appreciably more drought prone.

To counter water security threats, the OECD and BRIC (a Goldman Sachs inspired acronym for Brazil, Russia, India and China) countries are meant to spend $800 billion a year in 2015, 'a target likely to go unmet'. How much more does the developing world need to spend?[48]

Even so, the West's infrastructure can be something of a thin veneer, as we will see by the actual status of infrastructure in the USA and the UK's running hard to stand still in the face of various challenges. In fact, most of the West has only recently emerged from more basic times. In the USA, 80% of cities had sewerage by 1900, but only 4% of wastewater was treated. By 1960, 80% of wastewater was treated. Outside lavatories were the norm for working-class families in Britain up to the 1950s and sewerage was the exception in the Netherlands until the same time. In Belgium, just 20% of sewage was treated as recently as 1990, meaning that during their deliberations, all the sewage discharged by the European Union's environmental planners in Brussels flowed untreated into the River Senne.

Water is a finite resource and the shift and growth of humanity is putting these resources under increasing stress. While water is more plentiful than many believe, a scarcity of good management and political will can turn little local difficulties into a chronic crisis.

Ignorance is bliss...The dihyrdro monoxide story

Dihydrogen Monoxide, or Hydroxic Acid for short, is an odourless and colourless compound which can kill if accidentally inhaled. It is present in tumours of terminal cancer patients, is a major component of acid rain, as well as an industrial solvent and used in the production of chemical and biological weapons. Between 1989 and 2007, campaigns such as DHMO.org and the Dihydrogen Monoxide Research Division have sought to alert politicians and the public about this substance. In 2001, the office of a New Zealand Green Party MP was 'absolutely supportive of the campaign to ban this toxic substance', while six years later an MP for the country's National Party wrote to the Associate Minister of Health asking 'Does the Expert Advisory Committee on Drugs have a view on banning this drug?'[49] As many will have guessed, a popular name for this compound is water. There are – apparently – a few variants of this tale and suggestions that it will acquire urban myth status.[50] But, as we shall see, myths are often the bane of modern water management.

The story has legs. At the UN's COP16 climate change gathering in Cancun in December 2010, members of CFACT, a libertarian group circulated a 'Petition to Ban the Use of Dihydrogen Monoxide (DHMO)', claimed that 'almost every delegate that collegian students approached signed.'[51] As a wind-up, it was a signal success in demonstrating the divide that often exists between the people who want to set environmental and public health policy and those who in fact understand it. It is the difference between a throwaway quip and a quotation engraved in stone for millennia.

The DHMO story is a potent illustration of the sheer depth of ignorance that surrounds water management. Where scientific and engineering is most needed, it tends to be thinnest on the ground.

Shannon Information and the politics of fear

Why is the water sector so conservative when it so badly needs innovation across the board? The work of Claude Shannon and his popular muse Warren Weaver can explain how our perception of utilities works. When something is steady and un-newsworthy, we tend not to notice it. It may be the savannah shimmering as usual in the midday sun or a distant hum. It's only when things change that you become aware of them. Then a big cat comes hurtling towards you out of the heat haze or the hum either ceases or turns into a roar. Following their 'Mathematical Theory of Information' perturbations from an overlooked norm became known as 'Shannon Information'.[52] A strong bit of information will stand out from the background, and the stronger its message, the less information is actually needed. This to me explains modern expectations about water utility performance. When you turn a tap and out flows fresh, potable water, you don't notice the service because it is performing as you expect. The same goes when you flush the loo. If they don't perform to your expectations, you notice this and the service. Any deviation from the norm, which is assumed to be perfection, is seen as a near-catastrophic failure: the growing howl of popular anguish rising above the noise of time.

Yet fear and perfection can become self-indulgent and the costs can become absurd especially when involving media grabbing illness. Costs per cancer case avoided under the USA's Safe Drinking Water Act ranged from $0.5 million for ethylene dibromide to $4.3 billion for atrazine and alachor. In contrast, they found that an estimated $391 million per year in medium to large water treatment works would reduce giardia cases by 100,000 each year, working out at $782 to $978 per infection avoided, costs that would be comfortably exceeded by the medical and time costs.[53]

In a world of rolling news and a hunger for headlines, are you going to be responsible for one person getting cancer? If you are, spending $4.3 billion of other people's money may well feel like money well spent. A report urging the reduction in chlorine in drinking water announced that 240 million people living in the USA were exposed to tap water contaminated with some level of chlorination by products in 2001.[54] The numbers are then extrapolated so that exposure to these substances 'may place more than 700,000 pregnancies at increased risk of birth defects and miscarriages', which is far more dramatic (and in fact less quantifiable, since screening for early term miscarriages is rarely if ever carried out) than the 9,300 cancer cases the US Environmental Protection Agency believes are caused through chlorination each year. Should we do away with chlorination? It is one of the most effective sterilising agents and cost-effective alternatives remain a work in progress, especially where the water distribution systems are anything less than first rate.

Another challenge here is working out what is the optimum level of treatment. Ideally, all drinking water would be free from all contaminants. For utilities and regulators, reality means having to decide what is a practical level of purity. When researchers looked at the cost of dealing with lung cancer from arsenic, they found that removing arsenic from drinking water generated the best results. Lowering the concentration from 0.04 milligrams (mg) per litre in drinking water to 0.03 saved 0.88 lives per year at a cost of $10,000 per year. A further 0.88 lives are saved when the concentration falls to 0.02 mg per litre, but the cost of achieving this rose to $0.44 million per year, while a final reduction of 0.88 deaths by lowering the arsenic level to 0.01 mg per litre would cost $10.9 million.[55]

43

How much does it cost to save a life through new connections in developing countries? My estimate is $2,000–5,000 per life saved. Take the WHO estimate of people without access to safe water (884 million) and sanitation (2.6 billion) and put that against the 2.39 million who die from waterborne disease, or 1,090 non-connected people per death. I assume $50–200 per person (or $250–1,000 per connection, depending on where they live) and that new infrastructure will last 30 years (as many a Londoner knows, water networks still function, after a fashion, 150 years on) and this works out at $1,800 to 7,250 per life saved over that period.

Which offers the better value? The answer may appear simple, which explains why the funding flows towards the expensive options.

The value placed on avoiding a death carries immense responsibilities. Even so, it is hard to avoid the conclusion that developing and extending basic water and sanitation services can make a greater impact than fine tuning more advanced treatment options.

The complacency of some utilities

Services in England and Wales are a pointer as to how much room there can be for efficiency when utilities are progressively pressed and how hard it is to deliver true excellence and sustainability. The experience has many useful lessons for the rest of the world.

Many a water utility in the USA will proudly tell you that they are the best in the land, yet every survey of water and wastewater assessed the state of the nations' infrastructure at somewhere between 'D' and 'D-' where 'D' means dreadful.[56] One major player once told me that they classify their service offerings as for 'developed world' and 'developing world' standards and that the USA for operational reasons had to be included as a developing economy. The systems have degraded into near obsolescence in recent decades, framed by spending needs surveys which have been rolled back as no funding push emerges.[57] Part of the problem is the federal nature of the USA, whereby both policy and information is controlled at the state level, meaning that collating information that can be acted on is a slow and notably erratic process. The USA is unusual in its sheer diversity of small systems. When information and policy flows are localised, there can be a lack of overall direction. Contentedness is avoiding trouble, rather than appreciating your shortfalls in a broader context.

A poor infrastructure in the USA is often compensated for by using more chlorine, so people tend to use more bottled water. A vicious spiral can begin even though it may be better to spend the bottled water money on having pleasant tasting water flowing from their taps.

Good service does not guarantee good headlines, but shortfalls create better copy. When the then Yorkshire Water failed to guarantee that everybody would get water in the long hot summer of 1995 (which they in fact did) the entire board was forced to resign the next year while the company had to forgo £40 million in revenues from planned price increases and spend an extra £50 million in enhancing the security of its supplies. In the end, the company even had to change its name. Yet when Britain faced a drought back in 1976, people stoically waited by the standpipes when asked, and Dennis Howell, the specially appointed Minister of Drought, advised Britons to 'share a bath with a friend.' In the far-off days of avocado suites and sour industrial relations, putting up with setbacks was part of living in a nation embracing managed decline.

The weather might have been the same but the political climate had changed beyond measure. The 1950s to 1970s marked the apex of state control of England and Wales's water utilities, which were wholly privatised in 1989. The drought of 1995 made them tinder for the feeding frenzy of media savvy politics. When it was privatised, Yorkshire Water became a case study in the power of Shannon Information.

The pure absurdity of the 'crisis' was highlighted when it emerged that tankers were bringing in water to top up supplies. Would people have preferred that they stayed away and the pipes ran dry? Continuity of supplies is what matters, not the way the water arrives. At that point, people serving the network were allowed to work in their casual clothes, to avoid the prospect of being assaulted. Meanwhile, I was advised that Welsh Water was topping up reservoirs in South East Wales, but only by deploying tankers at night, buttressed by adverts in the local papers assuring customers in the area that their services were safe, which they were. The actual service delivery was identical, but the management of its delivery and the media were tellingly different.

What about telecoms, railways and electricity? Dropped mobile calls and poor or no signals are part of what is expected with a value-added service. After all, people do not buy designer clothes for fit or comfort; they buy them for the label. Railways in the UK are about poor service, they are meant to run late, be overcrowded and ever costlier. It is part of our propensity for putting first-class people into second-class lives. Power cuts invariably take place when you have been tapping away at something unusually compelling and lucid on your PC that has not been saved. This is a blip. In water, you don't get outages, you get outrages. This is one of the paradoxes with water, the more value-added the service, the more cheerfully lapses are accepted. But the political and media angst when a water utility comes up short is not a pretty sight, and this is why the water and wastewater utilities can be so risk adverse.

In whose interests are you being served? In truth, there exists a series of mismatches between expectations, delivery and reality. Ofwat, the water regulator in England and Wales certainly makes the companies perform (unlike our air terminals or rail services, for example), but it comes at a mighty cost.

Sometimes, you get the impression that when it comes to providing water and sewerage services a duality of expectations exists. Services are either meant to be consistently perfect or reliably poor, if they occur at all. All of this news must have generated some ironic amusement in those countries where if you have a piped water supply, you are expected to be grateful that it runs for the few hours a day that it does run, while most of the worse off are even denied the hope of household access.

While concentration has been focussed on the lack of infrastructure in the developing world, it is also evident that some developed economies are in some danger of finding that their water and sewerage networks are more decrepit and less developed as they imagine them to be, not least because of the under-developed nature of their political priorities.

Conclusions – a disconnected world

Water, when simple, is insipid,
inodorous, colourless and smooth.

Edmund Burke,
A Philosophical Enquiry into the
Origin of Our Idea of the Sublime and Beautiful

The words 'urban' and 'urbane' originally suggested elegance, refinement and culture, while the cheerful chaos of much urban reality today hints that matters may have moved on. For water and sanitation, geography and geology matters far more than it does for power and telecoms. The lie of the land affects the amount of pumping needed to get water to a treatment works and to its customers as well as sewage away from them. The soil affects the lifespan of underground assets, even more so when a city is built on sloping clay. Drive down some of London's residential roads at night and you will see gentle ripples upon their surface, not good news for pipes under pressure. The underlying rock will determine how much groundwater may be available and how much treatment is needed. Then there are the rivers; how fast they flow, what condition they are in and how much can be sustainably abstracted and discharged. Every town and city is unique and ever changing.

The developing world has been left behind

We face many problems seeking any form of proper access to urban water and sanitation. Perhaps we have been dealing with these for millennia, but recently have they reached the scale we are confronted with today. Despite sanitation systems serving the Indus Valley and Roman civilisations millennia ago, many areas across the world that once had fully functioning urban sewerage systems are now struggling to develop ones that can deal with today's demands, let alone the challenges of the near future.

While much has been spent expanding and upgrading basic water and sewerage services in high income countries meaning that universal coverage is rapidly becoming the norm for most urban dwellers in the 31 OECD countries, this has not been the case in most of the developing world. The sheer scale and pace of population growth and urbanisation has left national plans and international initiatives foundering in its wake. This has not been helped by the low priority given to these services by many of their governments. The number of urban dwellers without access to safe sanitation (as well as for water in Sub-Saharan Africa and South Asia) is actually rising at a time when it was meant to fall. At a time when waterborne disease is one of the leading causes of death worldwide, as well as rendering people too ill to work. We like to believe that we live in the modern world but there are places where we cannot provide the water and sewerage services our ancestors took for granted. Such a mismatch is unsustainable.

Universal access requires a universal sense of commitment

Providing everybody with safe water and sanitation should be quite straightforward, were it not for the fear that politicians and the public hold scientists and engineers in account and, likewise, the ability of lobbies to convert public anxieties into tools for their own advancement. We need to appreciate the balance between risk and returns when dealing with what lies in our drinking water, and how we perceive a utility's performance is affected by expectations and ownership. Developing services fit for our future purpose is a costly and long-term undertaking. As such, it rests uneasily with our current obsessions about instant news, sound bites and quick fixes. Arguments have to be developed to oblige policymakers and opinion formers to take the subject seriously.

So are we all doomed? Not at all, for as we will see, there are plenty of examples of good governance and innovation when it comes to making water management both universal and sustainable. For example, in 45 years, Singapore has gone from open defecation and shortages and standpipes, with rivers serving as sewers, to a comprehensive, universal and integrated water and sewage management system, ranging from desalination to wastewater recovery. Those years coincide with the country's independence and consistent political support for sustainable and affordable water management. The government of Singapore has appreciated the link between progressively developing these services and economic development and human well-being. Water has played a key role in the country's emergence as a truly developed economy.

To consider how this impasse has come about, a useful start would be to take a look at the water cycle and how we have modified it both directly and indirectly. The impact of the biggest indirect influence is already concerning water managers, but its real impact has yet to be felt: climate change.

---

[1] Simpson J A & Weiner E S C (1989) *The Oxford English Dictionary*, 2nd Edition, Volume XIX, Oxford University Press, Oxford, UK
[2] Simpson J (1988) *Touching the Void*, Jonathan Cape, London, UK
[3] Beer R (2004) *The Encyclopaedia of Tibetan Symbols and Motifs*, Serindia, Chicago, USA
[4] Little M et al (2001) Milk Consumption in African Pastoral Peoples in de Garine I and de Garine V (2001) *Drinking: Anthropological approaches*, Bergham Books, USA
[5] Goodall S K (2003) Rural-to-urban Migration and Urbanization in Leh, Ladakh. Mountain Research and Development, 24, 3 220–227
[6] WHO / UNICEF JMP (2010) *Progress on Drinking-water and Sanitation 2010 update*, WHO, Geneva, Switzerland
Arif G M (2000) Recent Rise in Poverty and Its Implications for Poor Households in Pakistan, *The Pakistan Development Review*, 39:4, 1153–1170
[7] Kleiner S M (1999) Water: an essential but overlooked nutrient, *Journal of the American Dietetic Association*, 99 (2); 200–206
[8] Thompson J P et al (2001) *Drawers of Water II: 30 years of change in domestic water use and environmental health in East Africa*, IIED, London, UK
[9] Shiklomanov I A (1999) World water resources and their use: A joint SHI / UNESCO product. UNESCO database accessible on http://webworld.unesco.org/water/ihp/db/shiklomanov/
[10] Pearce F (2006) *When the Rivers Run Dry*, Eden Project Books, London, UK
[11] Oldeman L R et al (1991) World Map of the Status of Human-induced Soil Degradation (GLASOD). 3 map sheets and explanatory note. UNEP, Nairobi, and ISRIC, Wageningen, The Netherlands.
[12] McKinsey (2009) *Charting Our Water Future: Economic frameworks to inform decision-making*, 2030 Water Resources Group / McKinsey, London, UK
Winpenny J et al (2010) *The wealth of waste: The economics of*

wastewater use in agriculture, FAO water reports 35, FAO Rome, Italy

[13] UNU-IWEH (2010) *Sanitation as a Key to Global Health: Voices from the Field*, UN University Institute for Water, Environment and Health, Hamilton, Canada

[14] WHO and UNICEF (2000) *Global Water Supply and Sanitation Assessment 2000 Report,* WHO / UNICEF, Geneva / New York

[15] McKenzie D & Ray I (2009) Urban water supply in India: status, reform options and possible lessons, *Water Policy* Vol 11 No 4 pp 442–460

[16] WELL 1998, *Guidance manual on water supply and sanitation programmes*, WEDC, Loughborough, UK

[17] WHO and UNICEF (2000) *Global Water Supply and Sanitation Assessment 2000 Report*, WHO / UNICEF, Geneva / New York

[18] *British Medical Journal* (2007) Medical Milestones Supplement, Brit Med J 344 Suppl 1

[19] WaterAid (2006) *Total sanitation in South Asia: the challenges ahead*, WaterAid, London, UK

[20] WSP-SA / MoUD (2008) Technology Options for Urban Sanitation in India. Water and Sanitation Program-South Asia, The World Bank, Delhi, India

[21] WHO / UNICEF (2006) *Meeting the MDG drinking water and sanitation target: the urban and rural challenge of the decade*, WHO, Geneva, Switzerland
WHO/UNICEF (2010) *Progress on sanitation and drinking-water*, WHO, Geneva, Switzerland

[22] Payen G (2011) Worldwide needs for safe drinking water are underestimated: billions of people are impacted, in Smets et al *Le Droit à l'eau potable et à l'assainissement, Sa mise en oeuvre en Europe*, Académie de l'Eau, p45-63, 2011.

[23] HDR (2006) *Human Development Report 2006. Beyond scarcity: Power, poverty and the global water crisis, UNDP*, New York, USA

[24] WSP-SA / MoUD (2008) Technology Options for Urban Sanitation in India. Water and Sanitation Program-South Asia, The World Bank, Delhi, India

[25] United Nations (2010) *World Urbanization Prospects: The 2009 Revision*, UN, New York

[26] UN-HABITAT (2003) *The Challenge of Slums: Global report on human settlements, 2003*, UN-Habitat / Earthscan, London, UK

[27] UN-HABITAT (2003) Improving the lives of 100 Million Slum Dwellers: Guide to Monitoring Target 11, UN-HABITAT, Nairobi, Kenya

[28] UN-HABITAT (2007) *Enhancing urban safety and security: Global report on human settlements 2007*. Earthscan, London, UK

[29] WHO / UNICEF JMP (2010) *Progress on Drinking-water and Sanitation 2010 update*, WHO, Geneva, Switzerland

Household sewerage access data was derived by the author from the UN JMP database of individual country survey data: http://www.wssinfo.org/documents-links/documents/?tx_displaycontroller[type]=country_files

[30] Neuwirth R (2010) The 21st Century Medieval City, The Long Now Seminar, San Francisco, USA, 10th June 2010

[31] United Nations (2010) *World Urbanization Prospects: The 2009 Revision*, UN, New York

[32] WHO (2003) *Emerging Issues in Water and Infectious Disease,* World Health Organization, Geneva, Switzerland

[33] WHO (2003) *Emerging Issues in Water and Infectious Disease,* World Health Organization, Geneva, Switzerland

[34] Krishna A (2010) *One Illness Away: Why People Become Poor and How They Escape Poverty*, OUP, Oxford, UK

[35] Howard G & Bartram J (2003) *Domestic Water Quantity, Service Level and Health,* WHO, Geneva, Switzerland

[36] Bartram J, Cairncross S (2010) Hygiene, Sanitation, and Water: Forgotten Foundations of Health, PLoS Med 7(11): e1000367. doi:10.1371/journal.pmed.1000367
Hunter P R, MacDonald A M, Carter R C (2010) Water Supply and Health, PLoS Med 7(11): e1000361. doi:10.1371/journal.pmed.1000361

[37] WHO (2003) *Emerging Issues in Water and Infectious Disease,* World Health Organization, Geneva, Switzerland
Bartram J, Cairncross S (2010) Hygiene, Sanitation, and Water: Forgotten Foundations of Health, PLoS Med 7(11): e1000367. doi:10.1371/journal.pmed.1000367
Kosek M, et al (2001) Cryptosporidiosis: an update, *The Lancet Infectious Diseases*, 1: 262–269.

[38] Bartram J, Cairncross S (2010) Hygiene, Sanitation, and Water: Forgotten Foundations of Health, PLoS Med 7(11): e1000367. doi:10.1371/journal.pmed.1000367

[39] Nriagu J O (1983) *Lead and Lead Poisoning in Antiquity*, New York, USA
Needleman L & Needleman D (1985) Lead poisoning and the decline of the Roman aristocracy, *Classical Views* 4,1: 63–94
Scarborough J (1984) The myth of lead poisoning among the Romans: a review essay, *Journal of the History of Medicine and Allied Sciences*, 39, 4: 469–475
Phillips C R III (1984) Old Wine in Old Lead Bottles: Nriagu on the Fall of Rome CW 78 (1984), 29–33

[40] Lessler M A (1988) Lead and Lead Poisoning from Antiquity to Modern Times, Ohio J Sci 88 (3): 78–84

[41] Lanphear B P et al (2000) Pediatric Lead Poisoning: Is There a Threshold? Public Health Reports 2000 (115); 521–529

[42] Hayes C R & Skubala N D (2010) Is there still a problem with lead

in drinking water in the European Union?, *Journal of water and Health*, 7(4), 569–580

[43] Diamond J (2004) *Collapse: How Societies Choose to fail or Succeed*, Viking, NY, USA

[44] Webster D L (2002) *The Fall of the Classic Maya,* Thames & Hudson, New York, USA
Hauh G H (2003) Climate and the collapse of Maya civilization, *Science*, March 14 2003, 299: 1731–5

[45] Weiss H & Bradley RS (2001) What Drives Societal Collapse?, *Science* (New York: American Association for the Advancement of Science) 291 (5504): 609–610
Kummu M (2009) Water management in Angkor: Human impacts on hydrology and sediment transportation *Journal of Environmental Management* Volume 90, Issue 3, March 2009, Pages 1413–1421
Evans D, et al (2007) A comprehensive archaeological map of the world's largest preindustrial settlement complex at Angkor, Cambodia, Proc Natl Acad Sci USA 104:14277–14282
Fletcher R, et al (2008) The water management network of Angkor, Cambodia, *Antiquity* 82:658–670
Penny D, Pottier C (2005) Hydrological history of the West Baray, Angkor, revealed through palynological analysis of sediments from the West Mebon, Bulletin de l'Ecole française d'Extrême-Orient 92:497–521

[46] Zhang S et al (2007) *Study on Water Tariff Reform and Income Impacts in China's Metropolitan Areas: The Case of Beijing*, The World Bank, Washington USA

[47] Vörösmarty C J et al (2010) Global threats to human water security and river biodiversity, *Nature* 467, 555–61

[48] Ashley R & Cashman A (2006) The impacts of change on the long-term future demand for water sector infrastructure, in *Infrastructure to 2030: Telecom, Land Transport, Water and Electricity*, OECD, Paris, France

[49] Gnad M (2007) MP tries to ban water. New Zealand Herald, 14th September 2007

[50] Ridley M (1997) Acid Test: Dihydrogen Monoxide: Now There's a Real Killer, *The Daily Telegraph*, 15th September 1997

[51] United Nations (2010) *World Urbanization Prospects: The 2009 Revision*, UN, New York

[52] Shannon C E & Weaver W (1949) *The Mathematical Theory of Information*, University of Illinois Press, Urbana, USA

[53] Congressional Budget Office (1995) *The Safe Drinking Water Act: A Case Study of an Unfunded Federal Mandate,* CBO, The Congress of the United States, Washington DC, USA

[54] Baumann J et al (2001) *Consider the Source: Farm runoff, chlorination byproducts, and human health,*The State PIRGs / Environmental Working Group, Washington DC, USA

[55] O'Ryan R & Sancha A M (2000) Controlling Hazardous Pollutants in a developing World Context: The Case of Arsenic in Chile, in Lehr JK & Lehr JK (2000) *Standard Handbook of Environmental Science, Health and Technology*, McGraw Hill Professional, NY, USA

[56] American Society of Civil Engineers (2001, 2005 & 2009) Report Card for America's Infrastructure

[57] US EPA (2009) 2007 Drinking Water Infrastructure Needs Survey & Assessment, Fourth Report to Congress, US EPA, Washington, USA

Chapter 2

The Water Cycle

Gatherings of the global great and good often include water in their communiqués these days. The World Economic Forum at Davos is one of the grandest as its prose shows, describing water as 'the gossamer that links the web of food, energy, climate, economic growth and human security challenges the world faces over the next two decades.'[1] Dew gathered on a spider's web perhaps, but water is better understood as something that flows in a vast cycle over land, sea and air in a solid, liquid and gaseous form.

Thinking about gatherings, there comes a point during most water conferences when a speaker is really excited to tell you that you might not realise it, but most water is in fact salt water and most fresh water is in fact ice. Many present will, at this point, catch up on their emails, make notes (about something else), smile politely or (especially if the 'gala dinner' was held the night before) bury their faces in their hands. Even so, some numbers about the 'Blue Planet' do matter:

Global water resources – where the water is and how it flows

	Volume (000 Km³)	% all water	% freshwater	Renewal period
Salt water (oceans)	1,338,000.00	96.5379%	–	2,500 years
Frozen				
– Glaciers & polar ice	24,064.10	1.7362%	68.708%	
– Permafrost	300.00	0.0216%	0.857%	10,000 years
Groundwater				
– Saline	12,870.00	0.9286%	–	1,400 years
– Predominantly freshwater	10,530.00	0.7597%	30.065%	1,400 years
Lakes				
– Salt	91.00	0.0066%	–	17 years
– Fresh	85.40	0.0062%	0.244%	17 years
Soil moisture	16.50	0.0012%	0.047%	1 year
Marshes and swamps	11.50	0.0008%	0.033%	5 years
River water	2.12	0.0002%	0.006%	16 days
In animals & plants	1.12	0.0001%	0.003%	Several hours
Atmospheric water	12.90	0.0009%	0.037%	8 days
Total – all water	1,385,984.64	100.0000%		
Total – freshwater	35,023.64	2.5270%	100.000%	

Putting these numbers into a broader perspective, most of the world's water is locked into slow moving cycles, (glacial cores have been dated to 800,000 years and drilling continues), while a fraction circulates pretty rapidly. These numbers are adapted from the work of Igor Shiklomanov whose *World Water Balance and Water Resources of the Earth* published in Leningrad in 1974 marked the start of our modern understanding about how, where and why water flows.[2] This work was updated and revised in 2003 for UNESCO's International Hydrographical Programme when Professor Shiklomanov led the State Hydrological Institute in St Petersburg.

'From cloud to coast' – how the water cycle works

Water is one of the major components of the biosphere, the life-supporting layer between the earth's inert rock and the near vacuum of space. It is quite a substantial veneer; if the earth was perfectly round and smooth, it would lie beneath 2.7 kilometres of water, including 70 metres of fresh water, 1.4 metres of which is readily available. Life is sustained within the biosphere by one open cycle (energy) and a series of closed cycles of water and nutrients (chiefly carbon, nitrogen, sulphur and phosphorous). Heat energy dissipates through the upper atmosphere and radiant energy comes in from the sun, more or less in balance. Apart from energy, the only things of substance to enter and leave the earth these days are satellites and spacecraft in one direction and meteors and dust in the other.

This was not always the case. The fact that water has been found on the moon and Mars and coats the surface of Jupiter's Europa shows it abounds in space. One plausible theory is that the primitive earth was showered with a great deal of ice over hundreds of millions of years. As the earth's crust cooled, a vast, supersaturated atmosphere of steam and water coalesced and fell in a primordial deluge. Whatever happened then, the seas formed and the water cycle cranked into action. The scale of this process remains awe-inspiring.

How water gets to where it does has mystified people for millennia: 'All the rivers run into the sea; yet the sea is not full; unto the place from whence the rivers come, thither they return again' as the Bible has it (Ecclesiastes 1:7). Authorities such as Aristotle assumed that rivers were supplied by condensation from vast caverns. Bernard Palisy, a French Huguenot potter, was unencumbered by the baggage of the Classics ('I have no other book than the sky and the earth') and was the first to appreciate the water cycle as he explained in a series of public lectures in Paris from 1575: 'When I had long and closely examined the source of the springs of natural fountains, and the place whence they could come, I finally understood that they could not come from or be produced by anything but rains...if the fountains and rivers came from the sea their waters would be salty...if the springs of fountains came from the sea, how could they dry out in summer?...Rainwater that falls on mountains, lands, and all places that slope toward rivers or fountains, do not get to them so very quickly...all springs are fed from the end of one winter to the next.'[3] By the eighteenth century the workings of the water cycle were broadly understood and the condensing caverns slipped into classical obscurity.

Water vapour circulates through the lower atmosphere, coalescing into clouds. Rainfall or precipitation takes place through a combination of cloud vapour density and gravity. Most rain falls back into the seas, but for our purposes, the water cycle begins with water being precipitated on to the land surface. On reaching the ground, it infiltrates the soil, runs off into the river system or is stored in glaciers via a number of processes. Water in the soil is either taken up by vegetation such as plants and trees where it is returned to the atmosphere as vapour excreted through transpiration, or it percolates through the soil. Once it has seeped through the soil, it either enters the river system or recharges aquifers (water bearing rock). From the aquifer, water emerges in springs, suffuses into the river system, is discharged into the sea through coastal springs or is stored in the rock. Some water from both river and ground water is also taken up by plants and in turn transpired, but most is discharged into the sea. Underneath the sun, water on the surface of the sea and lakes is heated up and evaporates. This is mainly from sea water. This vapour rises into the sky and accumulates as clouds, thus continuing the cycle, evaporation, the physical process accounting for 90% of atmospheric water and transpiration, the biological process the remainder.

The water cycle also breaks down into three geographic areas. Closed land regions (30 million Km^2) where water stays within the river basin until it evaporates. Here, there is a balance between rainfall and evaporation, of 9,000 Km^3 each year. Then there are the open land regions (119 million Km^2), the areas where water flows out of the region (known as exoreic run-off), generally to the sea. These areas have more rainfall (110,000 Km^3 each year) than evaporation (65,200 Km^3 each year) with the other 44,800 Km^3 flowing through rivers (42,600 Km^3) or as underground recharge and run-off (2,200 Km^3). Finally there are the oceans (361 million Km^2) with evaporation of 502,800 Km^3 and precipitation of just 458,000 Km^3 the difference between the two fuelling the surplus rainfall on land.

The longer water is held in a particular place, the smaller its active role in the water cycle. Its relative mobility means that on average 46 times more water falls as rain each year than is held in the atmosphere at any one time. Likewise, while the 10,530,000 Km^3 of fresh water in the groundwater system might appear to drown the 2,120 Km^3 of water held in the river system at any time, its residence time of 1,400 years works out as 7,520 Km^3 of water flowing through the groundwater system per year while a sixteen-day residence time explains why there is an average of 42,600 Km^3 flowing through the river system each year. Different estimates of inflow and outflow mean that the numbers here do not add up precisely, it is the relationship between volume and its flow that matters.

Our renewable water resources are the 42,600 Km^3 per annum that flow through the rivers, along with the renewable groundwater recharge of element of 2,200 Km^3 per annum. These can vary by 15–25% with for example drier than usual years worldwide in 1940–44, 1965–68 and 1977–79 and wetter years in 1926–27, 1949–52 and 1973–75. Just to confuse matters a little more, estimates for renewable water resources range from 33,500 Km^3 to 47,000 Km^3 per annum.[4]

When people say how little fresh water there is, they actually mean it is a large glass from a vast bottle. It is the flow of freshwater through the land system that matters, rather than how much might be in any place at any one time.

Water and life

All life upon the land depends on some minimal access to freshwater in a useable form. In contrast, most of the water cycle takes place independently of life. Without the first seas and ponds of primordial soup, the conditions for amino acids to meld into the building blocks of life would not have been there. Even if the earth was in fact colonised by extraterrestrial life or its building blocks, these too depended on water. Life would not have been able to evolve without water. Life would have been unable to colonise the land without rainfall and river run-offs.

While the water cycle would carry on if there was no life on earth, nature does influence it. Water vapour exhaled by plants encourages rainfall by creating the ideal conditions for cloud formation and precipitation. That is one of the principles behind planting trees to combat desertification. It appears odd that less rain falls upon each area of land than on the sea, but not all land masses produce much water vapour in the first place – water being scarce in deserts and the scope for evaporation limited where it lies locked up as ice.

Life takes place in a watery world – whole organisms can be seen as water containers, which retain the water needed for the biochemical reactions that sustain us, but only as far as we need it. Such cheerful reductionism would mirror the selfish gene philosophy best proposed in Richard Dawkins' 'the extended phenotype', where life is seen as expedient collectives of genes.[5] So, some water is held in plants and animals, mainly the former. Likewise, much more is held in the soil, which is itself a combination of the underlying rock, decomposing plant and animal life and the life that thrives on death. For plants and animals the water they carry is the water they need to survive. Except when they have to endure intermittent and scarce water supplies, there is little point in carrying excess water. A pot-bellied impala might not need to worry about where its next drink is coming from, but it would soon be very concerned about a lean and hungry leopard's remorseless advance.

Animals have largely evolved to do best when able to drink clean water. The exceptions are parasites, coprophytes and necrophytes, where the former are transmitted in drinking water which they have contaminated, with the latter two playing a leading role in enacting water and sanitation policy and spending plans as well as thriving on dung and death. Humanity's dependence on access to clean water reflects our vulnerability to contaminants, especially where people are brought up in relatively sterile conditions and have not built up a resistance to various common ailments. Nature's access to clean water is due to the water cycle acting as a vast treatment facility, the process of evaporation and precipitation being a global distillation unit – and a solar powered one at that.

Life on earth has coped perfectly well with the earth's freshwater resources for quite some time, including land based microbes for 1,000–1,200 million years, plants for 470–75 million years and amphibians emerging 100 million years later. Life has long thrived on earth and that ought to be the paramount principle when considering freshwater and humanity.

Unfair shares?

Of the 42,600 Km³ annual river flows, 20,426 Km³ is lost as surface run-off in floods (the South Asian monsoon, for example), while 7,774 Km³ flows through remote rivers, chiefly inaccessible parts of the Amazon and Congo, but also in northern Europe and America where rivers flow untapped into the Arctic Ocean. This leaves a net year-long, usable and accessible water input of approximately 14,100 Km³, which is sometimes known as basic run-off.[6] This compares with an annual abstraction of 3,829 km³ as estimated by the United Nation's Food & Agriculture Organization for 2000,[7] highlighting the scope for local imbalances between water availability and its need. It is evident that the element of the water cycle used by the human economy is not optimally managed. Much of the water abstracted is not put into productive use.

According to the Food & Agriculture Organization, 73.4% of water withdrawals in 2000 were from surface waters, against 19.0% from groundwater, 4.8% from drainage water, 2.4% from wastewater reuse and 0.3% via desalination. This means that 2,810 Km³ of surface water was withdrawn or 20% of accessible resources.

It is odd how much emphasis is placed on there being so little freely available fresh water. This viewpoint does not reflect the fact that in many areas (although of course, not all) that given half a chance, life on earth thrives under these conditions, and there is in functional terms enough water in the sky, streams and soil for the life that has evolved to live there.

Bemoaning the reality that so much water is salty and how much of the rest is locked up in glaciers is disingenuous at best, as this water would make little difference. Life on land needs the excess of precipitation over transpiration that comes from the seas, so a great majority of water needs to reside in the seas, and since rainfall starts with distillation, it does not matter whether seawater is salty or not. Anyway, when you consider what we pour into and drag out of the oceans, it is a blessing that they are so large. Glaciers and polar ice are not a lost resource; the icebergs and glaciers of the Arctic and Antarctic Circles wouldn't make a great difference (except to those living in low lying areas) if they melted and if the Arctic Sea was to become a sea again, the amount of rain-bearing clouds getting beyond the Arctic Circle would not alter matters greatly either. In crucial contrast, the glaciers of the Alps, Andes and, above all, the Himalayas are a vital resource, alternating between seasonal storage and melt, when their waters are discharged into streams, rivers and sea.

It is in fact extraordinary just how much freshwater there is, despite disasters such as the evisceration of the Aral Sea and Lake Chad. Four kilometres beneath the Vostock ice cap in Antarctica, there is a freshwater lake the size of Lake Baikal. Its existence was unknown and indeed unimagined until revealed by satellite images a few years ago. The water has not frozen because enough heat emerges from beneath the earth's crust while the ice above insulated the lake from what is the coldest place on earth: descending from -30°C in summer to -80°C in winter. This lake holds a major proportion of the earth's liquid fresh water. It is not a new resource; it has existed in utter isolation for a few million years, and we have done perfectly well without it.

Water does not flow steadily through history or across the lands. It fluctuates across time and space. Most continents have a wet season, April to June for Europe and South America, June to October in Asia, May to August in Africa and January to April in Oceania. Asia for example has 35.8% of total river run-off and 59.6% of the world's population while South America has 25.6% of global river run-off and 8.4% of the population.[8] While South America accounted for just 6.9% of water withdrawals, Asia accounted for 62.1%. In Asia, 80% of run-off takes place between May and October, accounting for much of the flood losses.

It is evident that the world's water resources are not a problem. It is their distribution and management in relation to current and future demand that challenges us, and only when unnatural demands are made of water resources do problems arise over where they are. Indeed, the very unnatural manner in which we treat our water resources today are a mighty pointer towards how it may be more sustainably managed tomorrow.

Managing the water cycle

Rain, as we have seen, is in essence distilled water; if the air is pure the rainwater is pure. But even in nature water does not simply circulate through air, ground and water intact. A plethora of other interactions take place during this cycle, adding to and subtracting from its chemical and biological loading, speeding up or slowing down its passage. In a natural system, sediments, minerals, nutrients and pathogens are constantly being discharged into surface waters. The pattern and impact of this discharge is affected by local circumstances, such as soil acidity (whether minerals are locked up or released by the soil), geomorphology (a fast-flowing river erodes, a slow-flowing river meanders and deposits) and seasons (river flow is lower and air and water temperatures higher during a dry season). The weather also has a powerful impact in many ways with surface water run-off from heavy rainfall bringing with it a loading of topsoil, excreta and pathogens, while shortening the time it takes for these to be flushed through a river system.

Soil and topsoil act as a water storage medium, retaining moisture that plants can abstract between rainfalls. The finer soil particles are, the more slowly water flows through the soil. This works both ways as soils with more clay content can absorb and retain more water, but sandy soils are less prone to surface run-off during heavy rainfall. If there is water-bearing rock underneath such as limestone, the surplus moisture seeps down. Groundwater usually undergoes filtration as it percolates down to the aquifers, the water-bearing rock. Under such conditions, it is often clear and sterile, simply interacting with the minerals present in the rocks. Here the bottled water blurb often hits the mark.

Wetlands act as localised water buffers, holding surface waters which would otherwise be flushed to the seas. Marshes, mangroves, estuaries and deltas function as settlement ponds, where solid wastes are deposited, shed as the flow slows, onwards to the sea, where the salt and sunlight in turn kill off more pathogens. The coastal shelf is fed by these run-offs, in contrast to the easing away of life's abundance towards the abyssal deep.

Outside the Arctic and Antarctic, glaciers act as a storage system. Restocked by snow during the winters, their summer melt can provide a relatively constant water supply at times when other water resources can be scarce. The glaciers of the Himalayas, Karakorum and Hindu Kush provide a constant flow of summertime water for some two billion people in China and South Asia. As anybody who has experienced a monsoon knows, there are other water flows, but glacial meltwater flows in a way which humanity can harvest.

The water cycle works, because terrestrial life has evolved to use it; wherever there is water there tends to be life. Life has also evolved to exploit the liquid and nutrient flows through the water cycle and after nearly half a billion years of evolutionary fine tuning, it can be said that the freshwater ecosystem services provided by nature are both extensive and advanced.

Water has a value

Since there would be no life without water, it clearly has some value. One way to consider the value of water is to look at the 'services' natural ecosystems provide in maintaining the water cycle. This is also known as 'natural capital' to contrast it with 'human capital', the value ascribed to our economic endeavours. Valuing such services may be an artificial process in that each system does more or less what its circumstances require and has a reasonable resilience, while nature does not exist to award medals and host ceremonies; it exists because it exists. However, as money talks to politicians and policymakers, it makes sense to theoretically monetise these services.

What does nature provide if you look at it as a utility? Vegetation acts as a water store, binding the soil against erosion; it is also a filtration system and, through transpiration, provides distillation services. Animals, by consuming vegetation maintain river flow. Microorganisms break down organic matter, especially in wastewater. Soil and silt act as water retainers and along with plants form a flood protection system. Water flow re-oxygenates water, while rivers and seas absorb wastewaters. Finally, natural re-growth performs habitat restoration services after floods, fires and drought and, in the past, has adapted to climate change.

How much do 'ecosystem services' contribute? Robert Costanza's paper 'The value of the world's ecosystem services and natural capital', published in 1997 in *Nature* was one of the first broad attempts to understand what nature contributes to what can be seen as environmental services that we otherwise take for granted.[9] His team evaluated the value of seventeen ecosystem services for sixteen global habitats ranging from the open ocean to tropical forests. Here, three services are of interest: water regulation, water supply and waste treatment. Some habitats such as ice, rock and desert do not provide these services and have no value, while others offer much of great value: lakes and rivers were seen to be worth $5,445 per hectare for water regulation, swamps and floodplains, $7,600 per hectare for water supply and marshes and mangroves $6,996 per hectare for waste treatment. Globally, they concluded that ecosystem services provided an average annual value of $1,115 billion for water regulation, $1,692 billion for water supply and $2,277 billion for waste treatment. Their total estimate for ecosystem services was $33 trillion against a global economy worth $18 trillion in 1994 dollars. I have some queries about the approach, as marine ecosystems are not seen as contributing to water supply (marine ecosystems account for $21 trillion out of the $33 trillion global total), yet – as we have already seen – they are a major source for rainfall onto land. Indeed, it would be interesting to value services such as rainfall in terms of material flows rather than hectares of habitat. Giving an illustrative $0.10 per M^3 value to rainwater as treated water running through the world's rivers, the 44,800 Km^3 of renewable resources would give an ecosystem services value of $4,480 billion each year. To mimic this by using the largest and most efficient reverse osmosis desalination systems available today would cost $0.50 per M^3 would cost $22,400 billion per annum, close to the our global economic output and at a mighty energy cost.

According to Truscost, an environmental consultancy, in an analysis for the United Nations called 'The Economics of Ecosystems and Biodiversity' (TEEB), the external costs or the price we would effectively be paying to compensate for the loss of natural water abstraction services was estimated at $1,226 billion in 2008 (2.04% of global GDP) and $4,702 billion for 2050 (2.92% of global GDP), with most of the 2008 costs being in Asia (57%), North America (14%), Europe (11%) and 18% for the rest of the world.[10] North America and Europe bear a relatively low burden here because of their extensive artificial water management systems. No figures for water pollution could be developed because of the lack of reliable data, especially outside the OECD.

Putting theory to practice, and translated into real estate, ecosystems do indeed have a value, and habitat preservation is a particularly attractive way to ensure future water resources. For example, New York City was faced with a long-term choice between spending $6–8 billion on new water treatment works and distribution infrastructure for protecting its traditional watershed in the Catskill Mountains.[11] In this case, they spent $1.5 billion buying up land surrounding its reservoirs and implementing management schemes to ensure that the integrity of their water inflows was maintained. As well as saving money (and longer-term operational costs), wildlife habitats have been preserved and enhanced at no extra cost to society, and public recreation improved and extended. In short, it is what management consultants call a 'win-win situation'. Closer to home in Mid Wales, Dŵr Cymru Welsh Water's Elan Valley Estate manages 18,000 hectares of uplands to ensure the quality of water Severn Trent supplies to Birmingham. Further north, the 9,300 hectare Lake Vyrnwy Estate preserves the catchment for the city of Liverpool, and its lease was sold in 2011 to North West England's United Utilities for £11 million.

Valuing ecosystem services is a work in progress. It has already become a useful tool for understanding the contribution that elements such as water and the water cycle can provide and even more usefully, by monetising such services, it speaks in a language that makes sense to those that look at the bottom line first and ask questions afterwards. Time will improve the models, but the message will remain the same, that both water and a healthy environment have a value.

Water and faith

In India, faith and water are mixed together in the Ganga (Ganges), the holiest of Hindu rivers. The Ganga can, at auspicious times and places cleanse the soul and wash away your sins. Sometimes water has special properties and the waters of the Ganga are perhaps unique in their self-cleansing abilities. The British East India Company used its water for journeys home as it stayed fresh for longer, while other studies found that the water could kill pathogens. Devendra Bhargava has spent decades looking at the river's quality and concluded that its waters can purify organic wastes at 15 to 25 times the rate noted in other rivers, and calculated that it can remove 60% of a river's BOD (biochemical oxygen demand) in 30–60 minutes, an order of magnitude above the absorptive capacity of most rivers.[12] No hard explanation for this property has yet been offered, but it fits in with the river's central role as a source of purity for Hinduism.

This is a most dramatic example of indirect potable reuse, which engineers such as Peter Cullen referred to as the 'magic mile', by which waste water is made fit for re-use by being mingled with river water.[13] The process is one that mixes biochemistry (how river water interacts with wastewater and to some extent can purify it, along with the sun's ultraviolet rays) and alchemy (things work best when out of our sight) and to people across the world, the 'magic mile' renews water in ways that human intervention cannot. Effluent discharged from a sewage treatment works can be and often is abstracted for drinking water use downstream. Along the River Thames, water can in theory undergo eleven such cycles of abstraction, treatment and distribution, use, collection and discharge before it reaches the sea. This capability clearly has a value.

Put a copy of the Bible through a word check and 'water' will appear 719 times. Water is almost as universal in faith as it is in life: 'I am Alpha and Omega, the beginning and the end. I will give unto him that is athirst of the fountain of the water of life freely' (Revelation, 21:6). As something of a sceptic, Philip Larkin shines an outsider's light on water and faith. In one of his most famous short lyric poems, 'Water', Larkin uses water as the central motif when considering how to construct a religion.

In Islam, Judaism, Christianity, Zoastrianism and Hinduism water is used in various acts of purification and ablution; for example in acts of initiation, preparation for prayer, and life's departure. 'Wash you, make you clean; put away the evil of your doings from before mine eyes; cease to do evil' (Isaiah 1:16). The Mandeans, who claim descent from Adam and Eve and revere John the Baptist, choose to live close to flowing water where they baptise themselves every Sunday in a ritual that combines purification and communion. As with Bon Po, other Animist beliefs such as Japan's Shinto abound with water spirits, often guarding water sources. Buddhism by contrast, avoids symbolism and water typically plays a peripheral role.

Such powers of purification have their limits when dealing with 400 million people's excreta, almost all of which is discharged untreated into the Ganga. Three thousand corpses floating down the river do not help and at the holy city of Varanasi, bacterial levels are between

120 and 3,000 times the official safe limit for bathing, depending on how close you are to the sewers and cremation grounds.[14] The 1985 Ganga Action Plan was meant to clean up the river by 1990, but bureaucratic and political delays and mismanagement have dogged the project. This remains at best a distant objective, since the 2006 official audit found that the Plan has only met 39% of its construction target, with Phase I of the Plan being completed thirteen years behind schedule. The River Conservation Authority that oversees the GAP only met eleven times between 1984 and 2006 although it did meet in 2008. In 2010, the National Ganga River Basin Authority was given another decade by India's Supreme Court to put the plan supported by $1 billion from the World Bank into effective action. Unfortunately, an exaggerated belief in the river's self-cleansing properties may again come into play.

The idea of the GAP was to ensure all the river waters would be of at least class B quality, or fit to bathe in by 1990. In 2003, 232 miles of the main Ganga Basin were class A and B and 1,201 miles were class C or worse, including 654 miles assessed as worse than E the lowest classification.[15] India's Central Pollution Control Board estimates that 8.25 million M^3 per day of effluent water is generated in the Ganga basin, against a current treatment capacity of 3.93 million M^3 per day. The GAP has been overwhelmed by population growth and urbanisation. Even the next phase of the GAP only aims to treat an additional 1.5 million M^3 per day of effluents. How many of these facilities work is another question since some were meant to run on biogas, but there was no biogas for them to run from. Others use electricity, despite the fact that power cuts are a daily occurrence, resulting in untreated sewage outflows. Finally, as the facilities only operate to secondary standard (offering a basic physical and chemical treatment to remove solids and the biochemical loading), they do little in the way of curbing faecal coliforms.

By giving water a spiritual dimension, humanity has traditionally revered water for its purity, which was a powerful incentive to maintain its quality. In India's Ganga, faith and a river's self-cleansing ability coalesced, but neither have been able to cope with the unnatural demands of today's urban humanity. Can modern humanity rediscover its primordial reverence for pure water?

How we modify the water cycle

Nature is in a constant state of flux; it always has its stresses and changes. In areas where water is abundant competition for those resources reflects this. Abhorring a vacuum life colonises areas of extreme water shortage as well, seeking to exploit niches that others are unable to endure. Thus camels and cacti would not have evolved unless there was a reason to and that was to be able to thrive where others cannot survive. The climate and the water cycle has changed in the past, causing change of habitat and ecosystems, sometimes on a colossal scale as during the ice ages. On a more local level, natural events – such as animal migration – can cause temporary water stress.

This brings us to humanity's impact on the water cycle. Freshwaters stay freshest when the cycle is in good flow, with new flows flushing though the system, constantly replenishing the system before oxygen levels fall and contamination rises too far as graphically illustrated down the Ganga. The more systems are compromised, the less resilience they enjoy. 'Stability needs diversity' is a catchphrase I conjured up as an undergraduate. It remains apt today; in a degraded system, one new change be it an invasive species or a new discharge will have a greater effect on the health of a river system than if a more intact system faced the same impact.

Michel Maybeck developed a working set of timescales for the global challenges and responses to continental rivers systems during what has been termed the Anthropocene era, when the human on the earth has become large enough to be marked in geological time.[16] There is some debate as to when this era starts, but there is a case for saying we are living in it.

The human impact on water resources over the past 2000 years

Year (AD)	0	1000	1800	1900	1950	2000
Environmental regulation	1	1	2	2	3	5
Sewerage / sewage treatment	1	1	1	2	3	4
Atmospheric pollution control	0	0	0	0	2	3
Artificial groundwater recharge	0	0	0	0	2	2
Land use	2	2	3	3	4	5
Urbanisation	1	1	2	3	4	5
River engineering	1	1	2	3	4	4
Mining	1	1	2	3	4	4
Atmospheric pollution	1	1	1	2	3	4
Agrochemicals	0	0	0	0	2	4

Key: Percentage of global area or global population affected

0 = 0%
1 = <0.1%
2 = 0.1–1%
3 = 1–10%
4 = 10–50%
5 = >50%

There are so many ways humanity interacts with water that it merits a shelf-full of books in itself and after years of increasingly intense and sophisticated research, these impacts are starting to be more fully appreciated. What follows is a brief overview, concentrating on some of the more pertinent issues.

Understanding water quality

Inland or surface water quality relates to what you would expect it to be where it is, influenced by the acidity of the rocks and soil, the rainfall, the shape and slope of waterways, whether the river-bed is hard or silty and the temperature. A simple definition of fresh water is that it contains less than 1000 milligrams per litre of dissolved solids such as metals and nutrients. Another indicator is turbidity, or cloudiness, caused by the presence of suspended solids in water. The European Union classifications range from 'Very Good' (IA) quality waters that have no appreciable indicators of human activities and are capable of supporting more sensitive species such as Brown Trout, to 'Poor' (III) quality waters that support a significantly degraded community of plant and animal species, and 'Bad' (IV) quality waters that (with the exception of some fungi and algae) are usually incapable of supporting life.

Biochemical oxygen demand (BOD_5) and chemical oxygen demand (COD_5) are the traditional biochemical and chemical water quality indicators and refer to the amount of dissolved oxygen in water consumed in test conditions over a period of five days. For BOD_5 this is by the microbiological oxidisation of biodegradable organic matter contained in the water, while in COD_5 it includes all the oxygen consumed by effluents. Water with a high BOD_5 or COD_5 will be degraded by its nutrient loading, holding less of the dissolved oxygen necessary for higher forms of life.

Measures such as BOD_5 and COD_5 are a useful indicator, but not much more. Since plants and animals are more concerned about the environment they live in rather than sets of numbers, interest has shifted to looking for indicator species in each body of water. That is the presence of plants and animals that you would expect to see in that habitat. In addition, in several families of invertebrates, better water quality results in a greater degree of species diversity. This has formed the basis of European Union water policy in recent years, which is forcing a fundamental rethink about the state of inland waters and how they are managed.

Ponds and lakes are also classified by their trophic state, or how much nutrients they contain. Oligotrophic lakes are often found in upland areas with acidic rocks. They have a low nutrient level along with a low amount of life, while the water is clear and pure. Mesotrophic lakes have medium nutrient levels, with plenty of plants and clear water. Eutrophic lakes have a high level of nutrients, with abundant life, but low dissolved oxygen levels. However many species of fish will be unable to live in a eutrophic habitat, which will also be vulnerable to algal blooms. Algal blooms denude the water below of oxygen and light, making them unsuitable for the life that once thrived there. Finally, there are hypereutrophic waters, where visibility is poor, and due to the lack of oxygen, dead zones may form below the surface.

Life thrives where there is fresh water, especially when it has good levels of dissolved oxygen. Pollutants and excess nutrients degrade water's capacity to sustain life.

Drugs – the human condition mirrored

Strange things happen when cycles get disrupted by drugs and stranger things still when the media decides it has picked up a good story.

An investigation by the Associated Press in the USA in 2008 ('Pharmaceuticals Found in Drinking Water') found that water supplies covering at least 46 million people in the USA contain a range of drugs.[17] There are seven billion prescriptions and purchases of non-prescription drugs in the USA each year. Do these findings reflect the state of America's psyche? Instead of California Dreaming, water serving the southern part of the state contained anti-epileptic and anti-anxiety drugs. Living in a city that never sleeps, New Yorkers' water boasted a mood stabiliser and a tranquiliser, alongside traces of heart medicine and oestrogen. It's not always sunny in Philadelphia, judging by the presence of treatments for high cholesterol levels, asthma, epilepsy, mental illness and heart problems. What the investigation highlighted was the general lack of testing for these drugs; there is no legal requirement for testing pharmaceuticals in water at the federal level and of the 62 major water supply systems approached by AP, 34 do not test for any drugs, while most of the others have looked at a limited range.

Montreal has the third largest sewage treatment plant in the world, which was completed in 1996, after 22 years of work. It is also one of the most basic plants serving a major city in the developed world, offering only primary treatment (screening and settling solids out and letting the rest flow away). With a quarter of its people taking antidepressants it was no surprise when the city's university recently found that the Brook Trout living downstream of the plant had elevated levels of Prozac and other drugs.[18] The municipality is currently considering spending C\$200 million on an ozone system to treat the effluents, but this will do little about drugs and heavy metals being discharged into a surprisingly cheerful ecosystem.

In 2001–04, Susan Jobling took a look at what happens when fish are exposed to treated sewage effluents in the 'magic mile'. She and her colleagues found that fish downstream from a wastewater treatment plant were changing sex due to low oestrogen concentrations. Endocrine disruption was subsequently found in fish populations across Europe, with male fish having ovarian tissue in their testes and in the more severe cases, lowered fertility.[19] In 2009 a five-year testing trial started in the UK after the

Drinking Water Inspectorate (DWI) reviewed the 500 most commonly used pharmaceuticals in the UK and prioritised twelve for monitoring, including, inevitably, one antidepressant. Traces of ten of these pharmaceuticals were found in sewage effluent and eight in rivers receiving these effluents with a survey of roach near 51 wastewater treatment works finding that 86% of the sites had intersex fish and at 20 of the sites, more than 30% of the fish being affected.[20] While the DWI's study into the presence of pharmaceuticals in drinking water found that sedatives and chemotherapy drugs at levels in drinking water were 300–30,000 times below recommended safety levels, these levels could harm a foetus.[21] The effect of long-term exposure to low doses is unknown – there may not be any effect, but such uncertainty is exactly where science struggles with human fears, no matter how faint their foundation.

In 2005 it transpired that scientists measuring the concentration of benzoylecgonine (a compound only found in the urine of cocaine users) measured the equivalent of 40,000 doses a day in Italy's Po Valley's river water.[22] This means that on average, one in 125 of the valley's 5 million inhabitants appears to be indulging in this habit every day, against official estimates that people living in the valley consume about 500 daily doses. Similar surveys in L'Albufera National Park near Valencia in Spain found cocaine, amphetamines, codeine, morphine and cannabis in the surface waters at between 0.06 and 78.78 nanograms per litre, while cocaine and ecstasy levels in Paris reach seasonal peaks during events such as Bastille Day.[23] Here, there is no apparent impact on aquatic life, but a reflection about assumptions and realities when it comes to understanding human nature.

Beyond the comic possibilities posed by the presence of all manner of pharmacopeia lies the simple fact that what we put into the water cycle, we may well need to take out. Discharging drugs into inland waterways may be a classic case of the law of unintended consequences.

Diversions – from dams to drainage

Fred Pearce's *When the Rivers Run Dry* is a pretty depressing read, being subtitled 'What will happen when our water runs out?'[24] He covers cases of poor management and over-abstraction ranging from salt build up in the over-irrigated soils of the Indus Valley, the slow death of Lake Chad and the Aral Sea, the sea of silt that is the Yellow River, the draining of the Marsh Arab's homeland to the strange death of the Rio Grande. His emphasis is on the impact of uncontrolled water use and the effects of our attempts to harness the water cycle. Worryingly, it is a useful primer about our problems.

The World Commission on Dams was set up to try to steer an increasingly acrimonious debate about the need for large dams and to balance their benefits (water storage and a source of hydro-electricity) against their costs (habitat loss, human displacement and the fact that some of them are being filled with silt alarmingly quickly).[25] We have to some extent become dependent on them, with 30–40% of irrigated land and 19% of global electricity coming via these dams including 70% of Europe's renewable energy and so act as an important element of minimising climate change.

Dams are a truly twentieth-century phenomenon: there were some 600 large dams in 1900, against 45,000 by

2000, their construction peaking between 1950 and 1990 with 5,000 built between 1970 and 1975 alone. While China had 22 dams in 1949, it has 22,000 by 2000, followed by the USA (6,390), India (4,000), then Spain and Japan. It appears that dam building has tailed off recently, except in India and to a lesser extent, China, Turkey South Korea and Japan. There are small dams aplenty as well. For example, Switzerland has 101,000 dams that are more than half a metre high.[26] Each of these is a way of regulating water flow and perhaps a local electricity supply, yet each is a barrier to the two-way process that is river flow and the flow of river life. By disrupting a river's flow and continuity, access to spawning sites for migratory fish such as salmon, trout, eel and sturgeon is impaired. Even small dams cause problems, as without a special passage they are impassable to most species of fish.

The downside to dams has been pretty apparent. The report notes that the World Resources Institute found that at least one large dam has modified 46% of the world's 106 primary watersheds, rising to 60–65% in Europe and the USA. Time erodes their utility, with dams and reservoirs losing their freshwater storage capacity at a rate of 0.5–1.0% every year due to sedimentation, which means that 12.5–25.0% of global fresh water storage capacity will be lost in the next 25 years, although anti-sedimentation measures are possible. Even so, China's Three Gorges are set to become the Three Tailbacks in a pretty short time – the Yellow River did not get its name lightly. Since dams and weirs hold sediment upstream, this makes the lower reaches of waterways vulnerable to erosion. With 26–48 million people officially displaced by dam building in India and China alone, dams have also been a driver towards urbanisation.

Rivers and streams may be modified in many other ways, especially through what is often termed hard engineering. Here, straightening watercourses makes them flow faster, preventing sediments from being deposited and indeed eroding river banks. In other cases, hard banks replace natural river banks, removing habitat and again encouraging faster river flow. New inputs can range from field and forest drainage schemes to foul and storm water outfalls. Water can be taken out (abstracted) formally at a treatment works or informally at any point. River beds may be dredged to keep channels open for shipping, especially where upriver activities have increased the sediment loading downriver.

Dams and diversions have been traditionally assessed on the basis of their intended benefits rather than their potential costs. As the latter are taken more fully into account, it is becoming increasingly clear that hard engineering approaches to watershed management are not necessarily the most efficient way of managing the water cycle.

Air pollution and rainfall

Normally, rainfall brings us the blessing of pure water, but if the air is polluted, it is the equivalent to contaminating drinking water after it has left the treatment works. Airborne particles combine with water vapour and indeed can help induce rainfall. Particulates may range from soot and sulphur (which brings acid rain) to dioxins, heavy metals and nuclear fallout.[27] In upland areas with high rainfall, these effects are magnified. To give one example, elevated levels of Strontium 90 occurred in British sheep due to rain-sped fallout from atom bomb tests across the world. These levels were falling by the 1970s due to test ban treaties, but after the Chernobyl disaster in 1986, restrictions were placed on upland areas where the rains brought a renewed burden of nuclear fallout. In 1986, 9,700 farms across Britain responsible for 4.25 million sheep were placed under these restrictions. Even in 2009, there were restrictions on 369 farms mainly in north Wales, covering 196,300 sheep[28] and in 2012, a consultation started to see if these restrictions could be eased.[29]

Nutrients, pollutants and surface water quality

Pollutants that affect inland waters include polyaromatic hydrates (from industry), polychlorinated biphenyls (from electrical transformers), which can bioaccumulate or build up in an organism's body to a dangerous degree, and pesticides.

The European Environmental Agency had a difficult gestation, a victim of power politics between Brussels and Strasbourg. Twenty years on and it has evolved into the world's most comprehensive regional gatherer and interpreter of environmental data, including an overarching synthesis, the 'European environment – state and outlook' which is published every five years. SOER 2010, as it is known has blossomed into a 52-part publication, with a synthesis on top.[30]

According to SOER 2010, excess nutrient inputs to agricultural land are the norm across Europe, causing eutrophication in inland waters, with a surplus nitrogen input ranging from 30 kilograms per hectare in much of Europe to 100 in Belgium, Denmark, Luxembourg and Germany and over 200 in the Netherlands. This is exacerbated where manure gets flushed into the river system from pastures, slurry and effluent stores. The turbidity (opacity) of waters rises where topsoil gets eroded through ploughing and rainfall. After a heavy rainfall, walking along two fields above our farm tells their story. One, regularly ploughed, runs in a thick torrent of red-brown water, while the other, a sheep pasture, has a smaller, clear flow, which carries on for several days. Many farmers across the world cannot afford fertilisers, let alone such excesses, but they do not enjoy the tax-funded largesse of the fiscal juggernaut that is Europe's Common Agricultural Policy. The European Environment Agency concludes that in 33% of the surface water stations across Europe where their trophic status is monitored, the water is either eutrophic or hypertrophic.

Pesticides are also a problem.[31] There are two main classes of pesticides: chlorinated hydrocarbons (such as Aldrin, DDT and Dieldrin) which are long-lived and capable of being concentrated up the food chain (this is called bioaccumulation) and organophosphates (such as Malathion and Parathion) which are short-lived and presumably degrade to 'harmless' end products, but whose long-term environmental impact is not yet known. As Kenneth Mellanby memorably noted in *Pesticides and Pollution* had Rachel Carson's *Silent Spring* kicked off with 'and no worms turn' rather than Keats' 'and no birds sing' it would not have had quite the same emotional impact. Tear ducts often get teased in the consciousness-raising process, but dry eyes are needed for honest drudgery and there is ample evidence that in rivers and lakes, short-term contamination after surface run-off can severely impact local invertebrate populations. Despite her occasional emotional overcharge and the opprobrium thrown at her by the chemical industry lobbies, Carson forced a fundamental rethink of pesticide usage in the 1960s and 1970s which impacts us to this day.

Inland waters reflect what flows and falls into them. Degrading a river's quality degrades its ability to deliver high quality water as well as being a high-quality habitat.

There goes the neighbourhood

Migration is an emotive issue in many places and certainly has an impact in surface waters. Europe has an inland waterways network extending for some 4,000 kilometres, interlinking rivers and canals. This lets animals and plants move from one habitat to another, and into areas where they have no natural predators. This is exacerbated by maritime trade bringing in alien life from other continents.

The Blue Danube has taken a bit of a beating in recent decades. In 2007 the aptly named killer shrimp was found in 93% of sampling sites and Asian clams at 90% of them. The water hyacinth can be a spectacularly invasive plant, capable of doubling its population in a week, especially where there is a high nutrient load. It can occupy the whole surface of a river or lake, choking native life and creating an idea habitat for malarial mosquitoes and parasitic flatworms. Fishing in Uganda's Lake Victoria can entail working your way through a deep reef of water hyacinth before open water is found. At the same time, they excel in absorbing nutrients and heavy metals and are used in wastewater treatment and can be harnessed as cattle feed and biofuels.

While invasive species are not usually a pressing concern in urban water management, the spread of water hyacinth demonstrates that what at one point is of academic concern can become a threat before people realise what is happening.

Flooding – rocks and hard places

Given the amount of money and effort made in harnessing water and controlling floods, it is remarkable how flood-prone many towns and cities have become. This is due to man-made factors; the modification of waterways, the modification of the ground and the modification of the climate.

The growth of hard-standing works on many levels especially in urban areas. It is the substitution of car parking for front gardens, and indeed the proliferation of car parks, it is roofs where the rainfall discharges into guttering and sewers rather than the surrounding grounds, and it is about urban sprawl, pavements, roads and paved open spaces punctuated by patches of grassland trampled hard so that rainfall cannot efficiently percolate down to the aquifers. Instead of much rainfall being absorbed into the soil, it runs off the surfaces straight into storm and combined sewers. Likewise, rivers have become more vulnerable to flooding as they have been engineered and odd as it may appear, 'hard' flood prevention and drainage schemes can make areas more vulnerable to flooding due to the removal of floodplains which acted as natural water absorbers, while narrower, straighter rivers overflow more easily. Exceptional rainfalls exacerbate these vulnerabilities. The 2007 flood at Severn Trent's Mythe water treatment works in Gloucestershire left 140,000 households without clean water for up to seventeen days. Such a flood is hardly exceptional in England now, with major flooding in Cornwall (Bocastle in 2004 and St Austell in 2010) and Cumbria (Cockermouth in 2009) in recent years. What have been referred to as 'once in a hundred years' events evidently have a habit of happening rather more often. Our stream, Nant Arberth ('Perilous Stream' in Welsh) discharges into the River Teifi near Llechryd, whose medieval bridge has been completely inundated several times in the past dozen years. The River Teifi and its surrounding fields and meadows have not changed that greatly in the past century, but the weather has.

The economic and human cost of flooding is rarely low. Yet blithe assumptions about the applicability of hard water engineering approaches have not been challenged until pretty recently. Concrete can only do so much in the face of climate change and the role and value of 'natural' flood defences needs to be reappraised.

Urban impacts

Does the health of the water cycle matter to people living in urban areas? It does, as it directly impacts the quality and quantity of their water, especially for towns and cities lower down rivers. Because urban areas often depend on river waters which have run through rural areas on the way, land management matters. Water resources are degraded by inappropriate farming practices, slash and burn 'development' of habitats and unsustainable forestry. Such practices lead to excess nutrient run-off, bacterial build up, pesticide contamination and increased silt loadings.

The proliferation of hard-standing in urban areas clearly affects the flow of rainfall into groundwater. Less rain reaching the aquifers means less groundwater recharge. Heavy rainfall can overload sewerage systems, especially when a combined foul (sewerage) and rainwater system operates. Sewer flooding is profoundly unpleasant and replacing combined systems is expensive.

Thinking about the footprints of cities, the greater the intensity of water abstraction, the more it reaches up and sometimes downstream and the same goes for the more sewage discharged, especially when untreated or partially treated. The longer and slower a river moves, the more these footprints matter. They also apply for industrial water and wastewater as well as for the municipalities themselves. Other impacts include heated wastewater and cooling water being discharged into waterways, most notably from power stations and the sheer energy intensity of pumping water and wastewater and treating it means that they have a further indirect impact on the water cycle.

Urban areas exacerbate the impacts humanity has on the water cycle, in effect creating an impact zone extending beyond each town and city. Within them, water flows need to be actively managed to avoid surface water run-off problems. They also depend on the quality of the waters that reach them. It is a complex and involved relationship with the surrounding land.

Groundwater depletion

Groundwaters have been taken for granted in many areas as a seemingly unlimited water resource. This has most dramatically been seen in countries such as Saudi Arabia where crops are harvested in deserts characterised by nearly non-existent rainfall. Groundwater sources which have been accumulated over millennia are being depleted in decades. If groundwater is over abstracted and it is near the sea, the adjoining aquifer will in all probability be filled with seawater as nature, abhors a vacuum. This is already a significant problem in many Gulf countries. Over-abstraction can also degrade groundwater quality as there is less water to dilute any pollutants being discharged into aquifers.

The status of Europe's waters

For inland waterways, a huge programme of assessing inland water status has been underway, focussed towards the Water Framework Directive, which aims to have Europe's surface waters attaining 'good ecological quality' by 2015 in theory and 2027 in reality. Daughter Directives have been passed seeking to attain the same for Europe's ground and marine waters.

How is Europe doing when it comes to good ecological status?[32] In 2007, the Netherlands concluded that 95% of their water bodies were heavily modified or artificial, against 2% in Ireland. In contrast only 21% of German rivers, mainly in less populated areas, are still in their natural state or only slightly or moderately altered. Such rivers may never retain their natural state, although intermediate measures can be taken. Overall, 40% were classified as being 'at risk' of failing the Directive, with 30% not at risk and 30% having a glorious let out called 'insufficient data'. 98% were 'at risk' in the Netherlands, along with 75% in the UK and 72% in Belgium. In contrast 25% of the UK's surface waters were not seen as being at risk against 0% in the Netherlands and 2–3% in Hungary, Slovenia and the Czech Republic. Insufficient data accounted for 65% of Spain's rivers but 9% for Belgium and 0% in the UK. It is hard to say quite how objective these assessments can be since when it comes to perceiving a country's environmental well-being, nature evidently has national characteristics.

Beyond Europe, assessing the health of inland waters is more anecdotal and episodic. There is little doubt that many other rivers, lakes and streams, especially in North America and Asia are just as modified. Many are more so and terrifyingly more so, at that.

Conclusions – water abounds, as does its abuse

As drop by drop the water falls
In vaults and catacombs

Alfred Lord Tennyson, *In Memoriam*

Water is almost ubiquitous in the biosphere, the part of the earth where life exists. It extends beyond life's reach, but life cannot extend beyond the watery realm. It flows across the land's surface in a swift cycle that provides a relatively constant supply of clean, well-oxygenated water. Its unusual properties are life enhancing in many ways.[33] Because it is lighter when solid, ices floats in water, allowing life to carry on underneath when a lake is partially frozen. You need to use a lot of energy to heat or cool it, helping mammals to control their body temperature. All of this is due to two quantum effects in water that cancel each other out, enabling its hydrogen-bond network to exist where they would usually not.

Until recently, human development was allied to its available water resources. Vulnerability to changes in water resources or its excessive use has been a constant contributory theme in the fall of past civilisations. This sense of vulnerability in turn has seen a respect for water resources, a respect compounded by the workings of the water cycle. The expedience of dealing with new demands for water caused by population growth, urbanisation and development has in turn impaired this intuitive understanding about the water cycle.

Instead, water has become regarded as a disposal medium, a 'magic mile' (if 'dilution is the solution to pollution' perhaps there really is a problem with the problem) and as a resource to be exploited regardless of the consequences of its mismanagement.[34] It can be argued that with a natural cycle of treatment and distillation, these are not material concerns, as water can be abstracted, treated and discharged irrespective of the water cycle – that humanity has transcended nature here. This overlooks the actual deterioration of water resources, the capability of natural systems to manage the flow of water and of our increased vulnerability to climatic perturbations as a result of the way the water cycle has been modified.

Water resources are diminishing in value at a time of increasing need. This is a disturbing inversion of the logic behind looking at nature as a provider of 'ecosystem services'. Replicating these services can be prohibitively expensive and carries with it the danger of further exacerbating the diminution of those very services they seek to replicate. Such additional costs would seriously compromise other spending priorities, such as providing basic access to water and sanitation services, as highlighted in the first chapter. It is time for water to be properly valued, both in nature and when managed by us.

[1] World Economic Forum (2009) The Bubble is Close to Bursting, World Economic Forum, Geneva, Switzerland
[2] Shiklomanov I H & Rodda J (2003) *World Water resources at the Beginning of the 21st Century*, Cambridge University Press, Cambridge, UK
Kalinin G P & Shiklomanov I H (1974) *World water balance and water resources of the earth*, USSR National Committee for IHN, Leningrad, USSR
[3] Denning D (2005) Born to trouble: Bernard Palisy and the Hydrologic Cycle, *Ground Water*, 43 (6) 969–972

4 Gleick P H ed. (1993) *Water in Crisis: A Guide to the World's Fresh Water Resources*, Oxford University Press, New York, USA

5 Dawkins R (1982) *The extended phenotype: The long march of the gene*, OUP, Oxford, UK

6 Postel S L, Gretchen C D & Ehrlich P R (1996) Human Appropriation of Renewable Fresh Water, *Science*, 271, 785–88, 9th February 1996

7 Comprehensive Assessment of Water Management in Agriculture (2007) *Water for Food, Water for Life: A Comprehensive Assessment of Water Management in Agriculture,* Earthscan, London, UK and International Water Management Institute, Colombo, Sri Lanka

8 UN Statistics Division, Department of Economic and Social Affairs (2009) World Population Prospects: The 2008 Revision, UN, New York, USA
Postel S L, Gretchen C D & Ehrlich P R (1996) Human Appropriation of Renewable Fresh Water, *Science*, 271, 785–88, 9th February 1996

9 Costanza R R et al (1997) The value of the world's ecosystem services and natural capital, *Nature*, 387(6230):255

10 PRI / UNEP FI (2010) Universal Ownership: Why environmental externalities matter to institutional investors, UNEP Finance Initiative, Geneva, Switzerland

11 Stapleton R M (1997) Protecting the Source: Land Conservation and the Future of America's Drinking Water, The Trust for Public Land, San Francisco, USA

12 Bhargava D S (1982) Purification power of the Ganges unmatched. L.S.T. Bull. 34. 52. Cited in Priyadarshi N (2009) Ganga river pollution in India – A brief report. Gangajal Blog 8 September 2009 (gangjal.org.in accessed 28th January 2011)
Bhargava D S (1983) Most Rapid BOD Assimilation in Ganga and Yamuna Rivers, *Journal of Environmental Engineering*. 109 (1) 174–188

13 Chartres C & Varma S (2010) *Out of Water: From Abundance to Scarcity and How to Solve the World's Water Problems. FT* Press, New Jersey, USA

14 *The Economist* (2008) Up to their necks in it, *The Economist*, 17th July 2008

15 Lacy S (2006) *Modelling the Efficacy of the Ganga Action Plan's Restoration of the Ganga River, India*, MSc Thesis, Natural Resources and Environment at The University of Michigan, USA

16 Meybeck M (2003) Global analysis of river systems: from Earth system controls to Anthropocene syndromes, *Philosophical Transactions of the Royal Society*, B, 2003, 358, 1935–1955

17 http://hosted.ap.org/specials/interactives/pharmawater_site/ accessed 10th February 2011

18 Gagnon C et al (2011) Distribution of antidepressants and their metabolites in brook trout exposed to municipal wastewaters before

and after ozone treatment – Evidence of biological effects, Chemosphere, In Press, Corrected Proof, Available online 5 January 2011

[19] Jobling S et al (2006) Predicted exposures to steroid estrogens in U.K. rivers correlate with widespread sexual disruption in wild fish populations, *Environmental Health Perspectives*, 114, 32–39
Jobling S et al (2002) Wild intersex roach (Rutilus rutilus) have reduced fertility, *Biological Reproduction*, 67: 515–524
Nolan M et al (2001) A histological description of intersexuality in the roach, *Journal of Fish Biology*, 58(1), 160–176

[20] Butwell T et al (2009) Removal of Endocrine Disrupting Chemicals – The National Demonstration Programme, Presentation at the CIWEM Water & Environment Annual Conference, 29th April 2009

[21] Watts & Crane Associates (2007) Desk based review of current knowledge on pharmaceuticals in drinking water and estimation of potential levels, Final Report to Defra, DWI, London, UK
Watts G (2008) How clean is your water? BMJ 2008;337:a237

[22] Zuccato E (2005) Cocaine in surface waters: a new evidence-based tool to monitor community drug abuse. Environmental Health: A Global Access Science Source 2005, 4:14 doi:10.1186/1476-069X-4-14

[23] Vazquez-Roig P et al (2010) SPE and LC-MS/MS determination of 14 illicit drugs in surface waters from the Natural Park of L'Albufera (València, Spain), *Analytical and Bioanalytical Chemistry*, 2010; 397 (7): 2851 DOI: 10.1007/s00216-010-3720-x
GWI (2011) Researchers' trip puts the fix on hard drugs, GWI Briefing, 10th February 2011, GWI, Oxford, UK

[24] Pearce F (2006) *When the rivers run dry: What will happen when out water runs out?,* Eden Project Books, London, UK

[25] WCD (2000) *Dams and Development: A new Framework for Decision-making. The report of the World Commission on Dams*, Earthscan, London, UK

[26] FOEN (2010) Core indicator: Structure of watercourses. Federal Office for the Environment. Bern, Switzerland. www.bafu.admin.ch/umwelt/indikatoren/08525/08586/index.html?lang=en (accessed 13th September 2010)

[27] Dalyell T (1975) Westminster Scene, *New Scientist*, 14th August 1975, 338

[28] Macalister T & Carter H (2009) Britain's farmers still restricted by Chernobyl nuclear fallout, *Guardian*, 12th May 2009

[29] Farming UK (2012) http://www.farminguk.com/news/Farmers-to-be-consulted-on-Chernobyl-restrictions_22460.html, accessed 10-1-2012

[30] EEA (2010) *The European Environment: State and Outlook 2010*, European Union, Luxemburg

[31] Carson R (1962) *Silent Spring*, Houghton Hifflin, Boston, USA
Mellanby K (1967) *Pesticides and Pollution,* Collins New Naturalist, London, UK

[32] EC (2007) First report on the implementation of the Water Framework Directive 2000/60/EC. Commission Staff Working Document (SEC(2007) 362 final). Brussels: European Commission (DG Environment). http:// ec.europa.eu/environment/water/water-framework/ implrep2007/pdf/sec_2007_0362_en.pdf

[33] Grossman L (2011) Water's quantum weirdness makes life possible, *New Scientist*, 2835, page 14, 22nd October 2011

[34] Chartres C & Varma S (2010) *Out of Water: From Abundance to Scarcity and How to Solve the World's Water Problems,* FT Press, New Jersey, USA

Chapter 3

Supply Management, Shortages and Scarcity

A brief history of urban water management

The development of aqueducts in Europe, irrigation systems in Asia and the Middle East, and water networks in Sri Lanka[1] from 300 BC – culminating in a comprehensive reservoir and canal network by 1186 AD – show how much effort was needed and, indeed, made in order to develop reliable water resources in early civilisations. Steven Solomon uses this as a way of comparing water despotism with water liberty, highlighting how at its extreme, many cities based on irrigation, aqueducts or reservoirs depended on captive labour to develop and maintain their supplies.

The agri-city could be relatively compact, an urban area fed by its surrounds. People living there practised intensive irrigation-based farming to feed the rest of the inhabitants. The land beyond could be somewhat detached from the city, being either arid (as in the Middle

East) or jungle (as in Angkor Wat and the Maya) and offering little beyond prospects of trade and game. In aqueduct cities (such as Rome and Pompeii, and well described in Robert Harris's page-turner novel *Pompeii*) water was brought to the city from supplies typically some miles distant, with most food growing taking place beyond the city walls. Such was the importance of Pompeii's harbour and strategic location that the cost and complexity of bringing water there could easily be justified. The World Health Organization notes that basic urban water distribution systems can also be a most effective way of spreading disease if and when they are contaminated and no treatment facilities exist.[2]

The river cities of Northern Europe and the North East of the USA and Canada were founded on water liberty. In lands of regular rains and rich earth, there was little need for special water schemes. The city could be fed by rivers and supplies augmented by groundwater, located in a rain-fed, agrarian landscape. While water was plenty here, its use and abuse was also plentiful, and soon water management and treatment was needed by these river cities, creating lush pastures for enterprise and entrepreneurs.

The world's first city-wide water treatment project was in developed by John Gibb in Paisley, Scotland in 1804.[3] Raw water was passed through a settling basin to take out solids and then through six feet of gravel followed by six feet of sand before use. This water was first intended for Gibb's textile bleaching plant with surplus water carted to customers and sold at half a penny per gallon (46p per M^3, or £14.74 per M^3 in 2011 money). Slow sand filtration was first used by the Egyptians, and small scale units were in use in early modern Europe. They were not quite sure what they were treating, but the system worked – showing how a little knowledge can go a long way.

Chlorine became the standard drinking water disinfectant as its effectiveness was quickly seen. In 1845 John Snow used it to control a cholera outbreak in London, although its first use for mains water did not occur until 1897. The ad hoc nature of the development and application of water treatment was most graphically seen during a cholera outbreak in Germany during 1892. Hamburg and Altona shared the same water source, but only the latter treated its water, via a slow sand unit before distribution. Hamburg suffered appreciably more than Altona did.[4]

Securing water supplies can be done simply through sweated labour or site selection. Treating water has seen a more complex evolution, until recently driven by concerns about aesthetics rather than disease – although the two had more in common than most realised.

Catchments and detachments, the art of supplying cities

There are a total 321 lakes, reservoirs and dams serving water supplies in England and a further 63 in Wales. The Elan Valley water scheme elegantly illustrates how the urban river-based water cycle operates. Rain from an 18,000 hectare catchment area in central Wales drains into four reservoirs; Caban Goch and Garreg Ddu (with a peak capacity of 35.53 million M^3), Pen-y-garreg (9.22 million M^3) and Craig Goch (6.06 million M^3) which were dammed between 1893 and 1904, and Claerwen (48.30

million M³) which was dammed during the period 1946 to 1952. Unlike earlier aqueduct projects, the workers were there of their own free will and accord, aided by the prospect of regular wages and on-the-job accommodation and food. Water goes into four pipes in a 118 Km aqueduct to Frankley Reservoir outside Birmingham which holds 0.53 million M^3 of water and was opened in 1904. The system has a capacity of 0.345 million M^3 per day and in normal conditions handles 0.320 million M^3 per day, dropping 52 metres over its length, meaning that the entire system is gravity fed. Normally, pumping is needed in order to transport water from one river basin to another. Frankley is a holding reservoir; its function is to ensure that there is an adequate supply of potable water to be released into the distribution mains where it may be held in a service reservoir to maintain water levels, while pressure is managed through a network of pumps and valves. The water leaves the mains down communication pipes of ever diminishing diameter until the water reaches the service connection at the consumer's property boundary. After this, it enters, via the customer supply pipe and serves the property, be it an apartment block, commercial, municipal or industrial building or even an Englishman's castle: the undivided household. The Elan reservoirs hold 310 days of water at normal delivery, while Frankley retains 40 hours of their flow.

This is about as good as it gets: a classic example of bulk water supply to a remote location. A secure supply of wholesome water that does not corrode pipes through acidity or lay down limescales across each household, while being, believe me, pleasant-tasting water. The good burghers of Birmingham are blessed by their Elan Valley water and, judging by the death rates prior to the aqueduct's opening, this was more than a matter of aesthetics; it was indeed life or death.

The flow from the Elan system provides water for 1.37 million people. It can either be delivered directly to customers from a distribution mains system (Elan water is soft, flavourless and low in minerals), or water from the River Severn is mixed in at the Frankley Water Treatment Works adjacent to the reservoir, where a granular activated carbon filtration system built in 2007–10 removes pesticides, herbicides and other organic contaminants from the river water, treating up to 0.240 million M^3 per day.

This reflects another reality, since as we have seen with the Thames and the Ganga, rivers are a product of their localities and what is poured into them from town and country alike. While river water serving some cities in Sweden is famously fit to drink, river water flowing through most urban areas usually needs at least some degree of treatment to make it potable. In contrast, most groundwater can usually be pumped straight from the aquifer to service reservoirs where it is held prior to release into the distribution mains without treatment. Groundwater contamination is now also becoming a problem.

The less densely populated parts of Severn Trent's service area, which covers the English Midlands and some of the Welsh March, are mainly served by the Severn and Trent Rivers. 61% of the drinking water supplied comes from rivers and lakes, 33% from groundwater and 6% from a mix of both sources.[5] Throughout the cycle, samples are taken for testing against a set of British,

European and world standards. In Severn Trent's case, during 2009, 117,240 tests were carried out at the water treatment works, 156,750 at service reservoirs and 277,041 at the customers' taps or supply zone. That is 551,211 tests for 7,580,000 people.[6]

A reliable and plentiful source of safe drinking water remains one of the greatest assets a city can enjoy when it comes to securing its long-term future. Transporting water over across a hundred kilometres or more can make sense when local supplies are inadequate, but such secure resources remain very much the exception.

A question of standards

The World Health Organization has been responsible for setting drinking water standards, starting with its 'International Standards for Drinking-Water' in 1958 and every decade or so since 1984 publishing 'Guidelines for Drinking-Water Quality'. The fourth edition emerging in 2011, along with extensive addendums[7] was 541 pages of main text, outlining all potential threats to be found in drinking water. In the introduction, the WHO points out that 'although the Guidelines describe a quality of water that is acceptable for lifelong consumption... should not be regarded as implying that the quality of drinking-water may be degraded to the recommended level.' This is a shopping list of what can and ought to be done, and behind this lies a debate about what is desirable versus what is affordable.

In 2000, the US Environmental Protection Agency proposed tightening the 1975 arsenic limit in drinking water from 50 parts per billion to five: half the WHO's 1993 guideline level which was adopted by the European Union in 1998. This was put forward as one of the last items in the Clinton administration having been modified to ten parts per billion and would affect 13 million people exposed to arsenic in their groundwater. When President George W. Bush took office later that year, he queried this change and it was kicked back to various subcommittees.

The debate ran round issues such as costs and benefits, as well as reflecting some tired partisan politics whereby Democrats are supposed to be 'pro' and Republicans 'anti' when it comes to environmental and public health standards. The EPA believed that the annual cost would be $181 million per annum, ranging from $0.86 to $32 per household for urban systems. In contrast, the American Water Works Association, while supporting the new standard, estimated it would need a capital cost of $5 billion and annual costs of $600 million for all consumers. The National Resources Defense Council[8] cited lifetime risks of dying from cancer by drinking two litres of water per day as running in a straight line from one in 1,000 at five parts of arsenic per billion to one in 100 at 50 parts per billion, while the World Health Organization believes that at ten parts per billion, there is a one in 1,667 chance of getting skin cancer, rather than dying from it.

In the end, the new ten parts per billion standard was passed in 2002 as the Bush Administration appreciated the sheer negative publicity their approach was generating, especially when it portrayed the USA yet again as a developing economy when it came to water standards. Debates continue, often spiced with arbitrary numbers, with stories about treatment for private wells costing $2,000 per year when domestic arsenic filter units in fact cost $160–260 with a three- to five-year warranty. A recent study by Yongsung Cho and his colleagues estimates that the cost of cutting arsenic to below the ten part per billion standard would be $230–2,006 per household, against a welfare value of $31–78 per household per annum where the water was above this limit.[9] The mass poisoning of rural villages via poorly located tubewells across Bangladesh in recent decades shows that at higher concentrations, arsenic is not a trivial matter.

What this episode illustrates is how data on economics and public health can be used in many ways when it comes to drinking water standards. This has a crucial bearing when it comes to deciding what money should be spent on, as we saw in the first chapter.

Engineering water

Degrémont, the water engineering arm of France's Suez Environnement, has published its Water Treatment Handbook in ten French editions since 1950 and seven English editions since 1958.[10] At nearly 2,000 pages it is a pretty useful starting point for those curious about water and wastewater engineering. While my well-thumbed sixth English edition from 1991 is still up to the task when it comes to appraising many 'new' technologies, the 450 or so pages of extra text in the latest edition has shifted the emphasis towards dealing with water stress and sustainable water management.

First there is the matter of ensuring a water source is potable or wholesome – that is, fit to drink. If it is not, then the requirement is to identify the problem and how that problem may be best treated. Water testing and treatment revolves round two concerns; is the water safe to drink and is it pleasant to drink.

The pleasant or 'soft' side of water quality is about aesthetics. It is about how water tastes, smells and looks. Unlike the health assessments, aesthetics is often a subjective art and so much of the testing is calibrated via panels of people considering water samples in a laboratory. Taste depends on place – water from an unfamiliar source may taste 'odd' simply because it is different. Other problems occur due to different levels of chlorine used to treat the water or bacteria and algae which can produce a 'musty' taste and smell. Activated carbon filtration or ozone at the treatment works usually deal with most taste and smell concerns, while the shift towards more advanced forms of water treatment lowers the need for chlorine in drinking water. Other concerns such as turbidity (how clear the water appears) and colour (iron making the water look brown and copper pipes in plumbsolvent areas leaving a blue residue) can be treated by a variety of approaches.

Unlike water safety, many of these problems may also occur within a household's pipes, where they may be corroded or water is not drawn for some hours or used for some time and the water reacts to metal or rubber fittings.

The role of chlorine in taste is fascinating as it has saved many millions of lives, but as we saw in chapter one, fears are now rising about the possibility of lives being impaired by high levels of exposure. Chlorine[11] does not deserve a place in the environmentalist's hall of shame, since too many ordinary people have benefitted too much for that to happen. It is perhaps an approach whose time is coming to an end, but it does show how things which were once regarded as beneficial can become objects of fear and loathing, stuck as they are between advances in human understanding on the one side and the ability to spread consumer concerns on the other.

Soft and hard water elegantly illustrate some of the balances between convenience and health. What do the gallops of Newbury, the great stud farms of Ireland, and Kentucky's Bluegrass Country have in common? They all boast calcium rich soil, which means that racehorses grazing these pastures have the strong bones demanded by flat and jump races. That calcium makes the water hard, or slightly alkaline. The post-gallop cuppa usually sees the kettle needing a good shake to wash out the lime scales, and washing clothes before modern detergents required a lot more soap, as suds don't form so readily. Deposits from hard water will in time affect boilers and block shower heads. Does this mean that slightly acidic soft water is invariably a good thing? Perhaps not, as soft water is plumbosolvent, meaning that lead will dissolve in it. The World Health Organization tightened its lead guideline from 50 µg/L (micrograms per litre of water) to 10 µg/L in 1992 with a compliance target in Europe from 2013. In areas with soft water, all utility pipes made of lead or with lead solder need to be replaced, along with domestic pipes and solder.

The 'hard' or public health side of water quality is about the legal concept of wholesome water, which means it complies with the various drinking water standards and failure to meet these can in some cases leave a utility vulnerable to legal action. In advanced water networks, health-related problems are rare. In England and Wales, overall compliance (that is, meeting all aesthetic and health related standards) has risen from 99.1% in 1989 to 99.9% by 2009. In 2006, Welsh Water was confronted with 231 cases of illness due to an outbreak of Cryptosporidium. Nobody was severely harmed, but with 71,000 people subject to a boil order (boil your tap water or use bottled water until further notice) it caused much inconvenience. No causal link between their facilities and the outbreak was found other than most of those affected being supplied by one reservoir. Outbreaks like this have their costs, with Welsh Water spending £0.5 million on an ultraviolet treatment unit for the network to ensure no further outbreaks, along with a £25 ex gratia payment to each customer affected.

Water can be purified by a great variety of processes, ranging from a single step (for example, to ensure no bacteria are present) to producing ultrapure water for specialist industrial processes. The more advanced an economy becomes, the purer its water. So, the water used to wash silicon wafers typically goes through seventeen separate treatment stages.

Treatments include coagulation and flocculation (a chemical process for the removal of particulates such as iron and aluminium), filtration (separating suspended particulate matter from a fluid), granular activated carbon filtration (for removing trace organic compounds), ozonation and UV (ozone gas and ultraviolet light respectively to remove pathogens) and chlorination (removal of bacteria). Other treatments may be applied for specific concerns such as excess acidity. More advanced treatments use higher degrees of filtration including ultra-filtration and nanofiltration, and reverse osmosis membranes and the membrane bioreactor much loved in desalination and water recovery. From here, new techniques are evolving in ion-exchange processes, especially where adopting substitutes for chlorine.

So, there are many tests and standards. In England and Wales, the Drinking Water Inspectorate has 44 principal 'hard' or measurable water quality standards, ranging from indicator values (a desirable standard) to UK and EU statutory limits. When new areas of concern emerge, such as drugs in water, the testing regimen widens. In the USA, the EPA has 90 statutory testing parameters, compared with 70 for Health Canada and 48 under the EU Drinking Water Directive.

Some of the substances that they tested for do not occur in waters (or are certainly not meant to) and there are concerns that too much effort is being placed on monitoring absent substances. Perhaps this is an appropriate use of the precautionary principle, but is it an appropriate use of limited resources when, despite the transgender fish mentioned in chapter two, formal drug monitoring remains very much the exception.

From the various offerings the River Thames has proffered over time, it is evident that potability lies in the gut of the beholder. Today, the tap water is almost certain to pass all applicable tests and, after the move towards advanced water treatment and the London Ring Mains, it is usually perfectly pleasant to drink.

Water standards are subject to many drivers, not the

least the ability to detect a pollutant in the first place as well as having a way of treating it. Some substances carried in water are essential for our health while it is often essential to have others taken out. The question of standards will constantly evolve as our understanding of the effects of various substances found in water advances.

Water scarcity and stress

While we live in a water-rich world, we don't always live where it abounds. As water supplies tend to be taken for granted, they are rarely factored in when populations shift and grow. Water stress is defined as 1,000 to 1,700 M^3 per person per annum and absolute scarcity as being below 1,000 M^3 per annum, with a recently adopted extreme scarcity category kicking in below 500 M^3 per annum.[12] According to the United Nations, water stress occurs when more than 10% of renewable freshwater resources are consumed. The European Environment Agency sees stress starting at 20% of renewable resources being abstracted annually rising to severe stress when abstraction exceeds 40%.

The pliant nature of such definitions can be seen when the European Environmental Agency regards Europe's water as being 'relatively abundant', with an 13% abstraction rate. A 30% abstraction rate in Spain masks abstraction rates of over 100% in Andalusia and Segura, where the excess has to be made up through river transfers and desalination, along with rivers failing to reach the sea during dry periods. In Denmark, the EEA reporting process states that all of Denmark's groundwater is vulnerable to over-abstraction and that it is vulnerable to temporary shortages, with 9.2% of renewable resources abstracted. Yet at 1,092 M^3 of water per capita in 2008, it is facing absolute scarcity, highlighting the difference between access to water and how it is used. Denmark uses its water resources with some care. Likewise, national pictures may not reflect regional realities. At a time when the Murray-Darling river basin was facing a severe drought and many cities were examining their water resilience, Australia boasted 22,957 M^3 per capita of renewable water resources in 2008, among the highest in the world. The sub-tropical forests of northern Australia are a long way from Melbourne and Sydney.

How water scarcity is defined: Water consumption as a percentage of run-off in a country

0–5%	No problems (Finland 1.9%)
5–10%	Temporary disturbances possible (Denmark 9.2%)
10–20%	Problematic – heavy investment needed (Australia 15%)
20–40%	Massive investment needed (Mexico 22%)
40–100%	Desalination and groundwater vital (Egypt 97%)
100%+	Dependent on non renewable resources (Saudi Arabia 164%)

The areas where economic development is growing the fastest are also those which are becoming the most water stressed. Here is the summary of research carried out by Veolia Environnement, looking at the need for water efficiency by comparing the share of global water and economic activity.[13]

Dry futures: Population and economic growth is set to take place in water scarce areas

% of renewable water withdrawn	% of global population		% of global GDP		GDP/Population	
	2010	2050	2010	2050	2010	2050
0–20%	46%	32%	59%	30%	1.28	0.94
20–40%	18%	16%	16%	25%	0.87	1.56
> 40%	36%	52%	22%	45%	0.61	0.87

While 32% of people living in OECD countries had no or low water stress in 2005[14] against 37% for the BRIC countries, 30% of those in the OECD are forecast to experience low or no stress by 2030 against just 20% for the BRICs. In contrast, severe stress in the OECD is forecast to rise slightly from 35% to 38%, while advancing in the BRICs from 56% to 62% during this period. So, the areas where the growth is meant to be taking place are also the most water stressed and becoming ever more so.

We tend to think that scarcity is something that recently cropped up, but it has a long history.[15] While they are not necessarily directly linked, it is fascinating to see that between 2005 and 2010, the majority of people were living in urban areas and subject to water shortages for the first time. The table below shows that scarcity was not recorded on a large scale until 1900 and extreme scarcity emerged from 1940.

Our scarce commons: People living with water shortage, total and percentage of the global population and by M3 per capita

Year	Total water short (million people)	Extreme < 500	Scarcity 500–1,000	Stress 1,000–1,700	All < 1,700
0	0	0%	0%	0%	0%
1000	5	0%	0%	2%	2%
1400	0	0%	0%	0%	0%
1700	6	0%	0%	1%	1%
1800	48	0%	0%	5%	5%
1900	131	0%	2%	7%	9%
1940	321	1%	4%	11%	16%
1960	571	1%	8%	10%	19%
1980	1,679	5%	11%	22%	38%
2005	3,247	10%	25%	15%	50%

Traditional adaptations to water shortage have been to build reservoirs and exploit groundwater resources along with more irrigation and trade in agricultural produce. The difficulty with these is that they only go so far and reservoirs tend to be built in wealthier and more water rich areas, for example in Europe rather than East Africa.

Water scarcity comes about when a finite resource meets an infinitely rising demand. Here, population growth is the problem, pure and simple, but many countries still seem to regard population size as a totem of power, rather than a driver of poverty and despair. Population pressure has moved from the occasional and periodic to the general and continual. Climate change brings in a new area of uncertainty with forecasts ranging between two and seven billion people facing absolute water scarcity by 2075, depending on which of seven climate scenarios you choose.[16] Pollution is also a challenge when water resources are so polluted that they cannot be used.

The Middle East offers an abundance of examples of the practicalities of living with water scarcity. Jordan's Amman-Zarga area has the one of lowest domestic water availabilities in the world, at 115 litres per person per day, but leakage and other abstraction lowers actual availability to 85 litres per day. Qatar's minimal rainfall means the country does not have any surface water resources, yet it has developed a farming system fed by groundwater. Withdrawal takes place at 188 million M^3 pa compared with a recharge of 50 million $M^{3.}$ and as a result the accumulated groundwater deficit during the period 1972–1995 was 994 million M^3 more than one third of the 1977 estimate of their total groundwater reserves. Saudi Arabia is another country that makes intense use of groundwater for agriculture. Drinking water mainly comes from desalination, but is not efficiently used, with desalinated water transported 200 Km from the coast to Riyadh,[17] where in 2006 31% was lost from leakage and the utility providing water for seven hours a day.

Sometimes water scarcity is a reflection of control within a region. So while Israel has an overall renewable water availability of 930 litres per person per day, resource allocations restrict use in the Palestinian territories to 57 litres per day. In the West Bank, people living in illegal settlements have an allocation of 110 million M^3 per annum compared with 50 million M^3 for the Palestinians.[18]

Definitions about water scarcity and water stress are somewhat interchangeable and often overlook the fact that life carries on even when it is supposed not to. It will have to since population growth, especially in water-short areas means that stress and scarcity are set to become the norm.

Supply management – keeping the customer satisfied

Supply management is a way of looking at water in terms of what the customer *wants* rather than *needs*. In reductionist terms it is about getting water to the tap, irrespective of circumstances or consequences. Water efficiency is therefore a nice idea but essentially peripheral to keeping customers happy. As a philosophy it is broadly unchallenged and it is also broadly unsustainable. A useful example is in England and Wales, where companies get a financial return on their infrastructure (called the 'regulatory asset base') and therefore the more infrastructure you own, the greater overall return you may enjoy. This largesse does not extend to tasks such as reducing leakage or installing water conservation systems at the customer end, but it does embrace desalination plants, reservoirs and other costly projects. Where water is metered, it favours the customer to use less, which penalises the utility, unless it can obviate the need for new spending.

You might not agree with him and his colleagues all the time, but Peter Gleick and his team at the Pacific Institute in California can resemble Thomas Huxley when he

116

declared himself to be Darwin's bulldog, ready and willing to take on all corners in the public debate over evolution. Looking for new water resources is part of the human condition; if there is a need, there needs to be a supply. This has much in common with oil, so it is little surprise that Peter Gleick's 'peak water' theory is becoming an outlier hit. Gleick coined the term 'peak water' to highlight the potential for regional water scarcity.[19] It stems from 'peak oil', the point when half of global oil resources have been consumed and production starts to level off. Peak oil may take place between 2012 and 2025, but we will only know when it does with hindsight and anyway, our dependence on oil may be diminished by a fundamental dash to gas.[20] Renewable water flows can only be used until they are over-abstracted. Since 1960, the Colorado River has flowed into its delta on just eleven occasions, with little or (more usually) no flow at all in the other 35 years, a case of peak water.

Groundwater offers plenty, but when overused, such as where more water is taken from an aquifer than its recharge rate, they become non-renewable resources. As the groundwater is depleted, abstraction will level out at the rate of natural recharge. However, if the aquifer has been contaminated by saline intrusion because of its overuse, further abstraction will require further treatment. Finally there is 'peak ecological water', highlighting where abstraction beyond a point reduces the value of ecological services provided by the water. Gleick points out that between 1900 and 1980, water consumption in the USA rose in line with GDP (at 2005 dollars), but since then, GDP has continued to rise while water consumption has been roughly steady. This may be because of more

117

efficient usage and allocation, but it may also be due to supplies hitting the buffers. Water is not running out, but there is only so much that can be beneficially used in any one place at any one time.

As with oil exploration, there always lies the hope of new resources. A decade or so ago, Earthwater Global was founded to mobilise and commercialise what they believed to be a new 'groundwater paradigm' – mega-watersheds. These were vast, undetected and untapped underground water resources, lying beneath conventional groundwaters, and which at various points could be tapped. There was a fairly vigorous debate in the water engineering community as to what this in fact meant and, at the last examination, the company (formerly earthwaterglobal.com) has become Geovesi (geovesi.com) 'a fully integrated groundwater exploration and project development and management company', concentrating on the Middle East and North Africa – but with no mention of the mega-watershed paradigm. It looks like being back to our usual resources and how they stand.

Water management has traditionally relied on the presumption that water can be channelled to where it is needed. As demands coalesce and water scarcity increases, a pure supply-led approach is likely to become increasingly untenable.

A complex evolution: London's water supplies

London's water history shows the private sector in all of its ragged glory – sometimes clearly a force for the good, sometimes a force otherwise. Without the private sector,

London's development would have been curtailed at crucial points in its history. London has been supplied by private water bearers for at least 515 years and by formal water conveyers for 430 years, in a relationship that continues to this day.

The evolution of water management in London is one of resource management and ad hoc entrepreneurs vying to supply an ever growing city and facing the consequences of the River Thames, their principal source, becoming unfit to drink. Reconciling an ad hoc lattice of supplies with changing realities would never be easy, especially as a city expands and expectations about service and public health become formalised. Let's take a look at how two such companies have evolved over time: the New River Water Co, a pioneering bulk supplier of water to the City, and The Barnet Water Co, a local supplier in the suburbs of Victorian north London.

Formal management of post-Roman London's water started in 1236 when the Corporation of London acquired abstraction rights for the Tyburn stream in the City and a conduit was built serving Cheapside in 1285. In 1442 another conduit was developed, bringing water from Paddington in Westminster's far west to the City. A Brotherhood of St Cristofer of the Waterbearers was formed in 1496 to regulate the carting of water round the City, enjoying 4,000 members a century later. In 1582, after eight years of lobbying the Corporation, Peter Morice's London Bridge pumping station was built and granted a 500-year lease to serve customers north of London Bridge. Morice's business prospered and in time became the London Bridge Water Company.

Hugh Myddleton[21] is probably Britain's most significant water entrepreneur, developing a clean water supply for London that has been in continual operation for nearly four centuries. After an attempt by Edmund Colthurst to dig a canal bringing fresh water from Hertfordshire to London ran into the ground in 1605, the Corporation appointed Myddleton to oversee a new scheme, and an Act was passed in 1606 to support it. Myddleton started the New River project in 1609, but ran out of funds a year later. By getting King James to provide 50% of the finance (in return for 50% of the profits) along with 29 'adventurers' for the New River. The project was restarted and its timing was fortuitous as the scheme's objectors were starting to smell blood. Having a royal investor overrode objections to the scheme and enabled the £18,525 project (£1.8 million in 2011 money[22]) to be completed by 1613. The New River Water Company was formally incorporated in 1619, the King owning 36 shares and the adventurers a further 36, with two of them holding one share and seven holding two. In 1631, King Charles sold Myddleton the Royal stake in return for an annual fee of £500. This turned out an injudicious move, given that the Company had started paying dividends on its 'benefitt profitt' in 1633. By 1640, a payment of £33 2s 8d per share was made, equivalent to £1,193 for the Royal stake. By 1695 the Company was among the three largest in England, along with the Bank of England and the East India Company, and its shares were the first to have been traded. The London Bridge Water Works Company was taken over by the New River Company in 1822 and its waterwheels no longer required. As London grew, New River flourished, supplying 270 million litres of water a day in 1886. It was estimated then to be worth £11 million (£628 million in 2011), a healthy long-term return on the original investment.

The Welsh Marches are studded with castles. Thanks to Sir Hugh Myddleton's roots in the March, these buildings resonate with echoes of the New River. In an ante room on the first floor of Powis Castle you'll find a framed New River Water Company share certificate signed by Myddleton. At Chirk Castle, home of Hugh's elder brother Thomas, there is a water pipe made from an eight-foot-long hollowed elm trunk.[23] Elm pipes were used to provide London's household supplies, with one tenth of households directly connected to the system by 1700. Their use was outlawed by the 1818 London Metropolitan Act, and they were replaced by the iron pipes that in turn are being replaced by plastic ones today.

Morice's waterwheel was burnt in 1666's Great Fire of London, and was soon rebuilt by his grandson, the system being expanded to five waterwheels by 1745. By 1700, twenty similar private suppliers or conduits were in operation, serving people along the banks of the Thames. It is prohibitively expensive to replicate water pipe networks, so each locality was served by what became known as a 'natural monopoly'. As London grew, so did the Thames's pollution loading, meaning that its water became progressively less potable but the conduit companies could compete against New River on price and locality. The most audacious of these water companies was the Grand Junction Water Company, founded in 1810, which originally took its water three feet from the Great Common Sewer's outlet by the Chelsea Hospital. Eventually they relocated to Kew where the Thames remained relatively clean. Others adopted water treatment. For example, the Chelsea Waterworks Company, incorporated in 1723 to serve Chelsea and Pimlico, where James Simpson built the world's first truly large-scale slow sand filter to improve its water in 1829, providing 85,000–115,000 litres of water a day.[24] Pressures on supplies forced some consolidation amongst the water conduit companies and by 1828 there were nine private water supply companies serving London.

The Public Health Act of 1848 ushered in the era of official concern about drinking water quality. The Metropolis Act of 1852 made water filtration compulsory for all water extracted from the Thames within five miles of St Paul's Cathedral and from 1855 companies could not extract water from the tidal river, which extends all the way to Teddington, nearly 30 Km upriver from London Bridge. A series of Acts passed between 1870 and 1894 gave municipalities the power to take over private water companies without recourse to Parliament. As a result, the London County Council took over the Metropolitan Board of Works in 1889 and in 1902 set up the Metropolitan Water Board for London. In 1903 the Metropolitan Water Board nationalised the New River Water Company and seven other water companies including Chelsea Waterworks for a total of £49 million (£2,796 million in 2011), testimony to the power and value of their piped networks. At that time, New River provided a quarter of London's water. Other private companies on London's expanding suburban fringe remained.

The Metropolitan Water Board encompassed Greater London, and in 1974 it was merged with twelve other local undertakings in the Thames Valley to create the Thames Water Authority, covering 13.8 million people for all sewerage services and 8.7 million for water today. The Authority was privatised in 1989 as Thames Water PLC which was acquired by RWE a German multi utility in 2001. RWE wanted to become a global water utility – and indeed at one point owned the largest water companies in the UK and USA, serving a total of 71 million people from Australia to Thailand – but they didn't appreciate the financial and regulatory challenges the sector presents. So they sold Thames Water in 2006 to Kemble Water, a financial consortium headed by Macquarrie Bank of Australia. Today, Thames Water is run as what is known as a 'cash cow' a utility milked for its dividends (£307 million in 2008–09 and £308 million in 2009–10) providing long-term returns for clients such as pension funds.

I once attended a board meeting of a water fund, specially convened at the Oak Room in the headquarters of the Metropolitan Water Board[25] in Islington, North London. Bill Alexander was retiring as Thames Water's CEO and wished to show us the jewel in the company's crown. The room was sumptuously panelled in dark oak, with a wealth of carving ranging from the Royal coat of arms to baskets of shellfish and fowl. It also boasts a section of the elm planking used for their mains water supplies. The counting hall and offices were empty, awaiting redevelopment as luxury flats as Thames Water sought to capitalise on its legacy. The room once formed part of The Waterhouse, the Company clerk's house at the New River Head in Islington which linked the New River from its source in Ware, Hertfordshire to Saddlers Wells,

Clerkenwell. From here it was taken to the City by water carriers crying 'fresh and fair New River Water! None of your pipe-sludge!', and later on through a network of (presumably sludge-free) elm pipes. The Waterhouse was expanded as the Company's headquarters in 1693, including the Oak Room built in the style of Grinling Gibbons. The Round Reservoir Pond was drained in 1913, and the Metropolitan Water Board demolished The Waterhouse and built its new headquarters in 1915–20 which incorporated the Oak Room and other historic fittings. Today, much of the New River has long been lost to urban development and its water now flows into the Thames Water Ring Main, completed in 1993. The catchment, however, still provides 8% of Thames Water's input. Another echo endures; its property portfolio built up round its Camberwell base was bought by London Merchant Securities in 1974 following the formation of the Thames Water Authority, and today this forms the core of Derwent London PLC's activities.

Even at the local level, water supplies can have a remarkably convoluted history, adding fresh layers to London's set of water stories. As well as what we know today as Thames Water, there were seven Statutory Water Companies operating within its boundaries by 1989. Each of these private water supply companies emerged through a combination of local circumstances and suppliers. Take the evolution of supplies to the once rural parish of Friern Barnet, now subsumed into north London. The water here was traditionally regarded as wholesome. In nearby Barnet Common a spring was found in 1652 and advertised as the 'Perfect Diurnall'; Samuel Pepys visited in 1664, drinking five glasses before setting home. Until

1866, the parish took its water from three brooks and some wells, with a conduit house noted as supplying people there in 1718. As the population grew, so did concerns about its water supplies.[26] In 1876, the New River Water Co erected a pumping station in the parish and from 1892, services were also provided by the Barnet District Gas and Water Company, which soon became the parish's leading water supplier. The company was founded in 1872, following the merger of local water and gas companies including The East Barnet Gas & Water Company (founded in 1866[27]) and renamed The Barnet Water Company in 1950 after Britain's gas services were nationalised. In 1960, it merged with Herts and Essex Water as Lee Valley Water. In 1987, the French water companies SAUR and Générale des Eaux (today Veolia Environnement) bought stakes in Lee Valley and the company was acquired outright by Générale des Eaux in 1988. Following clearance by the Monopolies and Mergers Commission in 1990, Générale des Eaux merged Lee Valley with Colne Valley Water and Rickmansworth Water, two neighbouring water companies to form Three Valleys Water.[28] This in turn was merged with North Surrey Water in 2000. Finally in 2009, Three Valleys Water was renamed Veolia Water Central. Three times, the company merged with other water companies, gas services came and went, and its name drifted from that of a locality to a corporate designation.

This complexity was mirrored across England and Wales. Water services, as we have seen, were often developed by entrepreneurs seeking to satisfy a demand. As long as this simply involved taking water from a convenient – and preferably pure – source and distributing it to their customers, this was a straightforward piece of business. As water supplies got contaminated and legislation emerged to address this, matters got ever costlier and more complex. That was the great driver towards the consolidation of companies and water bodies. In contrast, sewerage and sewage treatment services were typically developed by municipalities (albeit often with private finance) because of the greater capital costs involved and the challenges in making people pay for these services.

In 1910 there were 284 Statutory and other private water supply companies in England and Wales, along with a plethora of local water suppliers. At the time of the 1936 Public Health Act there were over 1,000 water entities, separately managed by 50 county councils, 150 borough councils, 600 urban and rural district councils, 33 joint water boards and 173 statutory water companies. This Act sought to encourage further amalgamations and takeovers. By 1963 there were just 29 statutory water companies and less than 200 water undertakings by 1974. Then in 1973 the Water Act merged all the municipal undertakings, including 1,300 sewerage authorities, into ten regional Water Authorities. This was an unintentional stroke of genius, as it means that England and Wales' water and sewerage entities are broadly grouped by their river basins, a move that anticipated European legislation by a quarter of a century. Like Thames Water, the other nine Water Authorities were privatised in 1989.

After two decades of inadequate investment and facing a raft of European environmental directives setting standards on bathing and drinking water and how urban wastewater should be treated and disposed of, by the mid-1980s it was evident that a once world-class infrastructure was falling behind European expectations. The water bills covered operating costs but not much capital spending, and the Water Authorities were not allowed to take on any significant debt. Spending on water, let alone sewerage and sewage treatment, is rarely a priority when governments have limited funds, and one of the reasons behind the full-scale privatisation (involving both company and assets) of the ten Water Authorities was to depoliticise infrastructure spending. Instead of competing for Ministers' attention, companies negotiate with an independent economic regulator (Ofwat, the Office for Water Services) as to what prices they are allowed to set. In return, they must carry out an agreed level of capital projects and meet service quality obligations which are monitored by two separate agencies, the Drinking Water Inspectorate for drinking water quality and the Environment Agency for inland and coastal water quality.

Has this worked? That depends entirely on your perspective. Failings in service delivery have been seen, but pale in comparison with those before them. Progress in leakage reduction and the adoption of metering have been erratic, but these also reflect conflicting priorities. On hard, quantifiable areas such as drinking water, bathing water and river water quality, a clear change for the better has been seen and European Union directives are being met. It has come at a mighty cost, with £84 billion being spent on upgrading and extending infrastructure between 1990 and 2010 compared with assumptions of a decade of catch-up work costing £19 billion back in 1989. That has meant an industry with cash reserves in 1989 (after a 'green dowry' from the Government) had net debt of £34 billion by 2010 and while operating spending efficiencies of 21% were attained, bills rose by 45% ahead of inflation during this time. I believe the experience has been a positive one, but the regulatory settlement needs to be overhauled to encourage companies to concentrate on efficiency and to be able to meet new challenges in a cost-effective manner. Karen Bakker's *An Uncooperative Commodity: Privatizing Water in England & Wales* reviews[29] the first decade from an academic perspective. We will return to this compelling subject later on.

Today just three of these ten 'Water PLCs' remain listed on the London Stock Exchange (Pennon / South West, United Utilities / North West and Severn Trent), four are held by financial investors, also known as the Private Equity sector (Anglian, Kelda / Yorkshire, Southern and Thames), one is owned by a Malaysian utility (Wessex Water), one by a Hong Kong infrastructure company (CKI) and one is run as a not-for-profit company (Glas Cymru / Dŵr Cymru Welsh Water). While there were 29 Statutory Water Companies in 1989, after a series of further mergers and acquisitions, twelve exist today: six held by international water companies, four by financial investors and two remain in local hands. Why so little of the sector is under British control is a reflection on international capital flows and the chronic conflict between the City's perceived short-termism and the sector's longer-term needs.

Other companies involved in water provision share complex histories. JP Morgan Chase, the American bank, started life as the Manhattan Company, founded in 1799 by Aaron Burr to provide water to Lower Manhattan. The company's charter allowed it to use its excess capital in any legal way, and Burr used this provision to establish New York's second bank which soon became the Company's principal activity. Water was indeed provided to over 2,000 customers using pipes made from hollowed pine logs until New York's municipal water system took over these operations in 1842. Until the Manhattan Company was acquired by Chase National Bank in 1955, its logo featured Oceanus, the Greek God of water. JP Morgan Chase's current logo was designed by Chermayeff & Geismar in 1961, the octagon being based on four wooden planks nailed together to form a water mains pipe.[30]

Compagnie Générale des Eaux was founded in 1853 to provide water to the city of Lyons in France. In the 1880s, it branched out to include sewerage contracts in France and water contracts internationally. Over the last 50 years, the company expanded into waste-to-energy, waste management and public transport. In 1987, the company gained the licence for France's second mobile telecoms network. I was a mobile communications analyst in the City at the time, and so I became involved in the business of water. The company started concentrating on its communications activities, as water did not attract the same sense of excitement and in 1998 it was renamed Vivendi. As attention turned to the possibilities of the Internet age, Vivendi Environnement was partially floated in 2000 (taking on a sizeable chunk of the parent company's debts) and was wholly deconsolidated in 2002. Two years later it was renamed Veolia

Environnement. In both JP Morgan Chase and Veolia Environnement's case, a single water contract has, over the centuries, spawned a global player with more than a thousand constituent companies.

Companies and municipalities have long sought to provide water for their cities. As understanding about the relationship between healthy water and public health has grown, these providers have been obliged to satisfy new standards, with companies moving on from pure entrepreneurs to regulated service providers.

An unwelcome reality – Climate change

If you are standing on a railway line and a train is trundling towards you, it may make more sense to step aside rather than arguing that the train is a figment of somebody else's imagination. Climate change is one of the constants actuating life on earth. We are living in an era between glaciations, previous ones which had made life all but unendurable in many areas. Even within written history there have been periods of drought, heat waves and mini ice ages, but all have been caused by natural phenomena. These are facts, as is the fact that we are living in another era of climate change and one which this time has almost certainly been caused by our activities. To understand how to manage the challenges facing our water services, you need to accept the fact that climate change is real, it is happening now and it affects the way the water cycle is working. Being brutal, climate change scepticism is about as valid as the attempts to deny the fact that the earth orbits the sun nearly 400 years ago. As far as water management goes, the expression 'global warming' misses the point, since climate change is what matters – the way patterns of rainfall have been changing.

Looking at the fevered pronouncements of the popular media, it would appear that 2009's *Climate Gate* exposed the 'myth of man-made global warming'.[31] Reality is rather cooler, as Fred Pearce observes in *The Climate Files* which is subtitled 'The battle for truth about global warming'. What the illegal hacking of the University of East Anglia's Climate Research Unit's emails revealed was the pressure non-political scientists were being put under from climate change denial activists cynically seeking to disrupt their work. Likewise, the Intergovernmental Panel on Climate Change has indeed been lax in allowing some unsubstantiated statements to slip through its review process, most notably the assertion that Himalayan glaciers could melt away by 2035.[32] But can one such paragraph undermine the 3,000 pages of research which has been broadly unchallenged since its publication in 2007?

Scepticism is part of the scientific process, especially when it is carried out to commonly accepted standards of scientific rigour. Climate change denial is about a closed mindset that holds the scientific method in ideological contempt. The denial community does not accept the need for accuracy. For example, Peter Glieck[33] highlights how the Heartland Institute (a conservative American think tank) 'disproved' climate change since 'records show conclusively that in April 2009, Arctic sea ice extent had indeed returned to and surpassed 1989 levels', while omitting to mention that the other eleven months showed exactly the opposite, as does the rest of the data from 1978 to 2010. It is evident that sceptical papers typically have little in the way of peer review and more than a few papers have unwittingly been co-opted to this cause. For example, one website called Popular Technology listed 500 'sceptical' papers as of the end of January 2010. Yet a brief examination finds that many of these 'papers' are merely opinion pieces in journals – that is, articles designed to stimulate debate – while others do not in fact refer to climate change per se or may be interpreted either way. Investigating the existence of various historic temperature cycles does not directly question human causality and indeed these cycles often cancel each other out as an explanation opposing human causality when placed alongside each other. I remember working with one international organisation a few years back where I was told they were banned from researching the impact of climate change on agriculture because the Bush Administration had threatened to withdraw their funding if they did. So, it is not as one-sided as it appears.

On the other side is the science. In 2004 Naomi Oreskes examined 928 peer-reviewed papers covering climate change published between 1993 and 2003.[34] She found that the 75% of these papers which were concerned with human causality all concluded that the human factor mattered, while the other 25% were looking at other aspects of climate change. None of these papers disputed the consensus. In 2009, William Anderegg, James Prall, Jacob Harold, and Stephen Schneider took a look at the records of 1,372 climate researchers and their views on anthropogenic climate change.[35] 97–98% of those surveyed concurred with the scientific consensus while the sceptical minority has less experience and a lower profile when it came to peer-reviewed research.

Cool appraisals versus heated opinions

The differences detected between the various reasons for a case for man-made climate change are small when compared with the differences posited for non-man-made climate change and indeed, it appears that the fallout between the various sceptic camps as to the 'real' cause of climate change are far greater than their objections to the IPCC. There is a mighty difference between people who are sceptical by nature (including several of my friends, so no offence is meant) and those who have adopted denial as a means of furthering policies or beliefs. Many of the sceptics have no scientific background and indeed appear to regard scientists and the scientific process with a patrician disdain. Presumably, when it comes to condensing caves, they continue to rely on Aristotle rather than East Anglia. I would ask such sceptics one thing: as you draw on your experience as an economist or classicist to reject scientific process in this debate, will you likewise elect to be operated on by an economist or a classicist rather than a surgeon in the event of an accident and if not, why not? The 'debate', like those over Heliocentrism and evolution, will doubtless drag on, spurred by humanity's fear of reality.

A divergent future is forecast

Three examples provide a flavour of the challenges ahead. When not fending off sundry conspiracy theorists, the University of East Anglia's Climatic Research Unit aims to disentangle historic weather data and to project how our climate may change. In 2007, Katherine Willett's team[36] found that between 1973 and 1999 a significant global increase in humidity in the atmosphere was primarily due to human activity. Such a trend is set to influence where rain falls and how heavily it falls. Warmer rivers and streams hold less dissolved oxygen and are more vulnerable to nutrient loadings. A team at the University of Maryland found that water temperatures in 20 out of 40 streams and rivers surveyed in the USA over 24 to 100 years had an average temperature rise of 0.009–0.077°C a year.[37] The rises were most pronounced in urbanising areas such as the Delaware River, but were also seen in pristine areas, pointing to climate change causing at least part of this change. In the uplands of northern England, the past two decades have seen a shift towards heavy winter rainfall and an 'almost complete absence of heavy summer rainfall', which was 'in marked contrast to the patterns seen in lowland areas.'[38]

Projections point to more rainfall and river run-off in high latitudes and the tropics and less in sub-tropical and other regions, which has meant an increase in dry areas. The shift towards more varied and extreme rainfall seen over the last century will increase and rising water temperatures will continue. The fourth assessment of IPCC, the Intergovernmental Panel on Climate Change concluded in 2007 that the likelihood of a causal relationship between human activities and climate change had moved from 'strong' (90% probability) in the 2001 Third Assessment to 'very strong' (95% probability).[39] It is not all bad, as the IPCC notes that forecasts for high latitudes water stress are 'very likely' to be reduced by climate change.[40] Predicted impacts of climate change on the water cycle include:

- More intense rainfall reduces groundwater recharge and increases surface run-off;
- More winter rainfall and less summer rainfall;
- More water loss by evaporation during the summer;
- Reduction in the recharge season for groundwater;
- Medium-term increase in summer glacial water run-off;
- Longer-term reduction in summer water from glacial sources.

Globally, the IPCC concludes that by 2050 twice the land area will be subject to reduced precipitation than increased precipitation as a consequence of climate change. Forecast changes affecting water management highlighted include:

- Exacerbation of pressures on water resources and their management;
- Higher frequency of extreme drought events;
- Inland waters will be warmer and therefore less able to absorb oxygen and less able to tolerate nutrient build-ups;
- Inland waters more vulnerable to over-abstraction and pollutant build-up;
- Increase in forestry productivity and agricultural productivity in certain areas, with increased water demand as a result;
- Decreased food security in Africa and Asia;
- Modification in patterns of tourism and outdoor recreational activities and changes in water demand where recreation increases or decreases;

- Increased need for storm sewerage systems and separate storm and foul water systems to deal with extreme rainfall.

Increasing temperatures intensify water demand, especially for irrigation agriculture and through human activities such as watering gardens and using swimming pools, and the need for cooling water for electric power and industrial plants. Changing seasonal patterns of precipitation also modifies demands for irrigation, particularly in regions with soils of low water-storage capacity.

Bringing matters round to water cycles and stresses, a team led Tajdarul Syed at the University of California using NASA's global monitoring systems recently made the first attempt to use full satellite monitoring to examine river run-offs worldwide.[41] They found an 18% rise in water discharge into oceans from rivers and glaciers from 1994–2006, or a 1.5% per annum rise in run-off. It looks like this rise is due to higher evaporation from the oceans, increasing the intensity of the water cycle. This may well be an exceptional period of rising river run-off, but something profound has happened in recent years and could be happening now. There are indications that more rain is falling in the tropics and the Arctic Circle and less in the semi arid regions where it is needed most. Indications of higher temperatures above the oceans mean faster evaporation (and more rain in general, but not necessarily on land) and more storms hitting the shores.

Adapting to a changing climate

The United Nations Framework Convention on Climate Change[42] oversees the UN's climate change programmes. Amongst these have been various attempts to estimate the cost of adapting to climate change. In 2007, they drew up a global estimate of $425–531 billion needing to be spent in the 23 years between 2008 and 2030, or $18–23 billion pa.[43] This scoping study looked at spending on reservoirs, desalination, water recovery and reuse and irrigation efficiency, finding the burden lay heavily with Africa and Asia: Africa ($131–138 billion), Asia ($238–288 billion), South America ($12–20 billion), Europe and North America ($37–86 billion) and Australia & New Zealand ($1–34 billion). The Convention then weighed the data to assess what measures related purely to climate change and concluded that $11 billion pa would be needed to 2030 ($2 billion pa in the developed world and $9 billion pa in the developing world).[44] This is an educated guess and these numbers will evolve as our understanding of the threats and the costs of adaptation evolve. The one lesson so far is that as we saw in the last chapter, the developed countries already have more resilient systems in place.

The human, environmental and economic costs do matter. The more the climate varies the more the damage as seen in the 1997–2000 El Niño-La Niña drought and floods in Kenya[45] cost the country 14% of its economic production during those three years. Likewise Mumbai in India received 94 cm of rain[46] during 26th July 2005, affecting 20 million people, killing 1,000 of them and causing losses of $1 billion. Such an inundation will test any storm drainage system at the best of times, let alone in a coastal city where there are tides to contend with.

The Stern Review, *On the economics of climate change*, is a useful synthesis of where we stand as to the potential effects of climate change and water management.[47] Globally, the forecast impact of climate change between 2050 and 2085 can be summarised in terms of what happens at each degree Celsius rise:

1°C 50 million affected by loss of Andean glaciers
2°C 20–30% decrease in rainfall in Southern Africa & the Mediterranean
10–20% increase in rainfall in Northern Europe & South Asia
3°C Southern Europe has serious droughts every 10 years
1–4 billion people exposed to water shortages (ME & Africa)
1–5 billion people face greater flood risk (S & E Asia)
4°C 30–50% decrease in rainfall in Southern Africa & the Mediterranean
5°C 750 million affected by loss of Himalayan glaciers

The current aim is to restrict the rise to two degrees Celsius. The current challenge is contending with those who do not see any need for any restrictions in emissions.

In the UK, looking at the range of forecasts at between the 33% and 66% probabilities, or where change is most reasonably likely to occur, on the low emissions scenario, summer rainfall is forecast to fall by 1 to 15% by 2025 and winter rainfall to rise by 3 to 13%, with somewhat larger changes for the higher emissions scenarios.[48]

How have things changed? In the real world, the debate ended a long time ago. Childhood memories of days of 'soft rain' have been replaced with deluges and water pouring off the nearby fields. It does remain a science in development, as more robust and locally applicable predictions are made about the anticipated effects and better time-series data emerges. On a personal level, it is alarmingly clear that the water sector has been at full stretch trying to cope with the day-to-day challenges of climate change for a decade. Bjørn Lomborg (a 'sceptical environmentalist' who accepts the reality of climate change) makes a very good point about adaptation being an immediate priority, one which sometimes gets overlooked.[49] Even so, it is somewhat disquieting to find that Sir Ian Byatt, the Director General of Ofwat from 1989 to 1999, and currently playing the same role in Scotland, is a climate change sceptic and one whose critiques have been found wanting.[50]

For most people involved with water management, it is extraordinary that there is still a 'debate' about climate change. The real challenge lies in the balance between responding to the more inevitable impacts of climate change and seeking to minimise their future impacts.

Conclusions – taken for granted?

If the world faces a crisis it will not be due to physical scarcities of water, but it will be due to sheer mismanagement of water

Professor Asit Biswas[51]

Supply management, the policy of keeping the customer satisfied first and asking questions later has, by and large, served humanity well for millennia. Unfortunately, in recent decades water scarcity has moved from being associated with droughts and dry seasons to a constant challenge for an increasing number of people. It has been driven chiefly by population growth, but also by urbanisation and its attendant demands, along with the degradation of some water resources and, most recently of all, climate change. While water resources are in one sense infinite because of recycling, they are also in another sense finite, in that only so much is available at any one time and not always where it is wanted most.

Campaigners against the private sector's involvement in water are typically defined by what they oppose, rather than what they propose to do about municipal incompetence, government indifference and the corrosive effects of corruption. There is a void between expectations and standards in some countries, notably in England and Wales, where anything less than perfection can result in a media-political feeding frenzy. Such attitudes are not helpful when seeking to extend service provision worldwide. If there is to be a universal right to water, it is a meaningless notion unless based on common standards and possibilities. Unfortunately, as the posturing over climate change has shown, we do not appear to have yet reached the point where enough people occupying positions of responsibility are willing, let alone able to act in a responsible manner when it comes to sustaining our water supplies.

Increasing demand is happening at a time when expectations about standards are also rising. Worries about water quality originally focussed on its appearance, and, by treating it, the water was made safe to drink as well. Since then, water treatment has shifted towards public health concerns and where these have generally been satisfied, back again to improving aesthetics. Technologies and techniques have evolved to meet these demands, as have those companies responsible for supplying drinking water. Private water companies originally concentrated on getting water to the customer, now the emphasis lies with the quality of the water that they get to their customers.

Dealing with concerns such as water scarcity and standards carries a cost, but as in the case of Saudi Arabia, where water is supplied at virtually no cost, it is being unsustainably managed and the costs of such mismanagement are going to be severe. Matters such as population growth and climate change are going to place these challenges in a much harsher light, whatever irrational optimists proclaim.

---

[1] De Silva S S (1988) *Reservoirs of Sri Lanka and their Fisheries*, Food and Agriculture Organization of the United Nations, Rome, Italy

[2] WHO (2003) *Emerging Issues in Water and Infectious Disease*, World Health Organization, Geneva, Switzerland

[3] Huismann L & Wood E (1974) *Slow sand filtration*, World Health Organization, Geneva, Switzerland

[4] WHO (2003) *Emerging Issues in Water and Infectious Disease*, World Health Organization, Geneva, Switzerland

[5] Severn Trent Water (2008) *Water Resources Management Plan 2009 – Volume 1: Appendices*, Severn Trent Plc, Birmingham, UK

[6] DWI (2010) *Drinking water 2009: Central region of England*, DWI, London, UK

[7] World Health Organization (2008) *Guidelines for Drinking-water Quality, Third Edition*, incorporating the first and second addenda, WHO, Geneva, Switzerland

8   http://www.nrdc.org/water/drinking/qarsenic.asp
    US EPA (2001) Fact Sheet: Drinking Water Standard for Arsenic,
    EPA 815-F-00-015, January 2001
9   Cho Y K, W Easter & Konishi Y (2010) Economic evaluation of the
    new U.S. arsenic standard for drinking water: A disaggregate
    approach, Water Resour. Res., 46, W10527,
    doi:10.1029/2009WR008269.
10 Degrémont (2007) *Water Treatment Handbook*, 7th English edition,
    Lavoisier, Paris France
    Degrémont (1991) *Water Treatment Handbook*, 6th English edition,
    Lavoisier, Paris France
11 WHO (2003) *Emerging Issues in Water and Infectious Disease*,
    World Health Organization, Geneva, Switzerland
12 Falkenmark M & Lindh G (1976) *Water for a starving world*,
    Westview Press, Boulder, Co, USA
    Falkenmark M & Lindh G (1993) *Water and economic development*,
    in Gleick P H ed. Water in Crisis: 80–91. Oxford University Press,
    New York, USA
13 Veolia Water (2011) Finding the Blue Path for A Sustainable
    Economy, A White paper by Veolia Water. Veolia Water, Chicago,
    USA
14 OECD (2008) OECD Environmental Outlook to 2030, OECD,
    Paris, France
15 Matti Kummu et al (2010) Is physical water scarcity a new
    phenomenon? Global assessment of water shortage over the last
    two millennia, Environ. Res. Lett. 5 034006
16 Falkenmark M et al (2007) On the Verge of a New Water Scarcity:
    A Call for Good Governance and Human Ingenuity, SIWI Policy
    Brief, SIWI, Stockholm, Sweden
17 Al-Musallam L (2007) Urban Water Sector Restructuring in Saudi
    Arabia, Presentation at the GWI Conference, Barcelona, April 2007
18 Friends of the Earth, Middle East (2005) *Crossing the Jordan:
    Concept Document to Rehabilitate, Promote Prosperity and Help
    Bring Peace to the Lower Jordan Valley*, FOE ME, Amman, Jordan
    and Bethlehem & Tel Aviv, Israel.
19 Gleick P H & Palaniappan M (2010) Peak water: Conceptual and
    practical limits to freshwater withdrawal and use, Proceedings of
    the National Academy of Sciences, published online before print
    May 24, 2010; doi: 10.1073/pnas.1004812107
20 Helm D (2011) Look to gas for our future and then our ingenuity,
    Prospect, April 2011, 45–48.
21 Barty-King H (1992) *Water; The Book*, Quiller Press, London, UK
22 Using the National Archives converter –
    http://www.nationalarchives.gov.uk/currency/results.asp#mid
23 Adshead D (2010) A craftie Fox's hollow log at Chirk Castle,
    National Trust arts, buildings collections bulletin, April 2010,
    National trust, Swindon, UK

24 Simpson J (1838) Filtration of Thames Water at the Chelsea Waterworks, *The Civil Engineer and Architect's Journal*, 1838, 15 p 392

25 http://www.britishlistedbuildings.co.uk/en-369261-new-river-head-revetment-of-old-inner-or
http://www.britishlistedbuildings.co.uk/en-369253-new-river-head-headquarters-173-177-isli

26 Baker T F T & Pugh R B ed. (1976) *South Mimms: Public services. A History of the County of Middlesex: Volume 5: Hendon, Kingsbury, Great Stanmore, Little Stanmore, Edmonton Enfield, Monken Hadley, South Mimms, Tottenham*, Victoria County History, London, UK
Baker T F T & Elrington C R ed. (1980) *Friern Barnet: Public services. A History of the County of Middlesex: Volume 6: Friern Barnet, Finchley, Hornsey with Highgate*, Victoria County History, London, UK

27 Hansard (1866) Local and Personal Acts, HC Deb, 10 August 1866 vol 184 c2166

28 Monopolies and Mergers Commission (1990) General Utilities PLC, The Colne Valley Water Company and Rickmansworth Water Company: A report on the proposed merger, MMC, London, UK

29 Severn Trent (2010) *Changing Course: Delivering a sustainable future for the water Industry in England and Wales*, Severn Trent Water Ltd., Birmingham, UK
Bakker K (2003) *An Uncooperative Commodity: Privatizing Water in England and Wales*, OUP, Oxford, UK

30 Tett G (2009) *Fool's Gold: How The Bold Dream Of A Small Tribe At J.P. Morgan Was Corrupted By Wall Street And Unleashed A Catastrophe*, Simon & Schuster, New York, USA
JP Morgan Chase & Co (2008) *The History of JP Morgan Chase & Co*

31 Pearce F (2010) *The Climate Files*, Guardian Books, London, UK

32 IPPC (2010) IPCC statement on the melting of Himalayan glaciers, IPPC, World Meterological Organisation, Geneva, Switzerland

33 Gleick P (2011) Misrepresenting Climate Science: Cherry-Picking Data to Hide the Disappearance of Arctic Ice, http://www.huffingtonpost.com/peter-h-gleick/misrepresenting-climate-s_b_819367.html

34 Oreskes N (2004) The Scientific Consensus on Climate Change, *Science*, Vol. 306 no. 5702 p. 1686, 3rd December

35 Anderegg W R L, Prall J W, Harold J, & Schneider S H (2010) Expert credibility in climate change, Proceedings of the National Academy of Sciences of the USA, June 21, 2010, doi: 10.1073/pnas.1003187107

36 Willett K M et al (2007) Attribution of observed surface humidity changes to human influence, *Nature* 449, 710-712 (11 October 2007)

37 Kaushal S S et al (2010) Rising stream and river temperatures in the United States, *Frontiers in Ecology and the Environment* 8: 461–466

38 Burt T P & Ferranti E J S (2011) Changing patterns of heavy rainfall in upland areas: a case study from northern England, *International Journal of Climatology*, n/a. doi: 10.1002/joc.2287

39 IPCC (2007) Climate Change 2007: Fourth Assessment Report

40 IPCC (2008) Climate change and water: IPCC Technical Paper IV

41 Syed T H et al (2010) Satellite-based global-ocean mass balance estimates of interannual variability and emerging trends in continental freshwater discharge, www.pnas.org/cgi/doi/10.1073/pnas.1003292107

42 Bates B C, Kundzewicz Z W, Wu S and Palutikof J P, eds. (2008) Climate Change and Water, Technical Paper of the Intergovernmental Panel on Climate Change, IPCC Secretariat, Geneva, Switzerland

43 Kirshen P (2007) Adaptation Options and Costs in Water Supply, A report to the UNFCCC Financial and Technical Support Division, http://unfccc.int/cooperation_and_support/financial_mechanism/financial_mechanism_gef/items/4054.php

44 Parry M et al (2009) Assessing the Costs of Adaptation to Climate Change: A Review of the UNFCCC and Other Recent Estimates, International Institute for Environment and Development and Grantham Institute for Climate Change, London, UK

45 Mogaka H et al (2006) Climate Variability and Water Resource Degradation in Kenya, World Bank Working Paper No 69. World Bank, Washington DC, USA

46 Danilenko A, Dickson E & Jacoben M (2010) Climate change and urban water utilities: challenges & opportunities, Water Working Notes No 24, The World Bank, Washington DC, USA

47 Stern N (2007) The Economics of Climate Change: The Stern Review, Cabinet Office – HM Treasury, London, UK

48 Defra (2009) *UK Climate Change Projections 2009 – planning for our future climate*, Defra, London, UK

49 Lomborg B (2001) *The sceptical environmentalist: Measuring the Real State of the World*, Cambridge University Press, Cambridge, UK

50 Byatt I C R et al (2006) The Stern Review 'OXONIA Papers': A critique, *World Economics*, 7, 2, 145-151, April–June 2006
Stern N (1996) Climate change: A reply to Byatt et al, *World Economics*, 7, 2, 153-157, April–June 2006

51 Biswas A K (2006) Water – managing a precious resource, *Pan IIT Technology Review Magazine*, 1 (3)

Chapter 4

Sewerage and Sewage Treatment

The effluent society

What comes in goes out. How much sewage we produce and how often will be affected by your size, age and health, along with how much you eat and what you eat. The average person usually voids 135–270 grams of faeces each day (35–70 grams dry weight), equivalent to 88–177 million dry tonnes of excreta worldwide every year.[1] Vegetarians generate more than omnivores, while in rural Africa, people on a very high-fibre diet happily shift stools weighing 400 grams (100 dry) every day.

Urine gets overlooked – which is odd, considering that we pass one to two litres of it every day, including an average of 5.3 kg of nitrogen, phosphorous and potassium each year, compared with just 1.0 kg of these nutrients in our faeces. With a total of 50–70 grams of dissolved dry solids passed each day, our urine actually contains more solid material than faeces.[2] These nutrients can go a long way to providing the fertilisers needed for food production.

There are many facets to faeces. You don't see the dry stuff, except in labs where people examine its composition and how and why people excrete what they do. Sewage generally refers to municipal wastewater, which is – or ought to be – conveyed to a wastewater treatment works via the sewerage network, from where, in theory, it is treated and disposed. Industrial wastewater is often called industrial effluent, but the terms can be interchangeable. Where there is a sewerage system and sewage treatment, urban sewage normally ends up as wastewater and then as sewage sludge. Domestic sewage is the term for household wastewater, which ranges from flushed lavatories (black water) to bath water (grey water). Wastewater contains 350–1,200 milligrams of solids per litre (250–850 mg/l dissolved solids and 100–350 mg/l suspended solids or 99.96–99.88% water) – the former in developed economies where lavatory waste is diluted by dishwashers and power showers, and the latter in developing economies where urban water is more sparingly used.[3] So wastewater is strongest in densely inhabited cities with a low penetration of white goods. Wastewater also gets concentrated where grey water gets recycled.[4] 70% of sewage solids are organic matter, 65% of this being proteins, 25% carbohydrates and 10% fats. The other 30% is inorganic material such as grit, metals and salts, especially where a sewer network combines household wastewater with road run-off. Half of those proteins are, in fact, bacteria.

Sludge from sewage treatment is typically 3% solids by dry weight. Members of the House of Lords memorably

146

likened sewage sludge to having 'the consistency of thin semolina' back in 1991.[5] Since transporting foetid water is costly as well as hazardous, it is usually dewatered before being taken away for disposal or recovery. Gravity thickening, or letting the sludge solids settle down takes the solid content up to 5–6% and dewatering by pushing the sludge through belt presses and the such-like gets the dry content up to 16–25%, depending on how hard you squeeze. It can be made drier, but the processes are costly and energy intensive. To give an idea of how much water gets squeezed out, consider Cairo. Each day, its Gabal El Asfar sewage treatment work takes in 600,000 m³ of wastewater. After treating and digesting (the biogas provides for 60% of the plant's energy needs) and pressing, 120 tonnes of 21% solid sludge are taken away as fertiliser.

Sludge is pretty powerful stuff. In addition to nutrient loadings and the bacteria that make up our gut 'flora', 50 of so infections can be transmitted through faeces, ranging from viruses and bacilli to parasites such as hookworms and tapeworms. Somebody with cholera can excrete ten million million vibrios each day, while we can unknowingly lay thousands of worm eggs every day.[6] This is what you might call 'the paradox of poo'; we cannot live without a healthy gut flora, but coming into contact with our excreta can be deadly. Microorganisms abound in sewage[7] with a billion faecal coliforms per 100 millilitres of raw sewage, along with thousands of salmonella, enteric viruses, helminths and giardia cysts. Later on, we will see what sewage treatment can do to some of these.

A brown glaze on the blue planet

When it comes to scatology and British humour it is usually the more prurient, the better. I am, by family reputation, someone with several decades of expertise in plopping inappropriate nuggets of such information onto the dinner table. So, one more heave: how much of the stuff is there? Take those 35–70 grams per person per day. That works out as 1.17–2.34 litres of sewage sludge (3% solids by weight) per day or 426–852 litres every year. According to the United Nations there were 2,837 million people living in urban areas in 2000 and these people could have generated 1,209-2,417 million m^3 of sewage sludge each year or 1.21–2.42 Km^3. When it comes to wastewater, the volumes rise dramatically, with people connected to sewage networks discharging 100–343 litres every day. This works out as 104 to 355 cubic kilometres every year.

The first systematic effort to map urban areas (24,000 urban areas with more than 5,000 people) was carried out by GRUMP, the Global Rural Urban Mapping Project, led by Columbia University's Center for International Earth Science Information Network. In 2005, they worked out that 3% of the earth's land surface was urban in 2000 – or 4.47 million Km^2. Putting these together, sewage sludge, if spread evenly across all urban areas, would lie 0.27–0.54 mm deep or the thickness of four sheets of office printer paper – in effect, a brown glaze. Not much, but urban areas today are much lower in density compared with the cities of old. Wastewater would create a deluge 5–16 cm thick. People living in slum areas would not be surprised by such numbers. Just to jolly things along, the world's urban population rose by an estimated 649 million over the following decade, or another sheet of paper.

Why the faecal fracas?

There is a lot of social conditioning attached to faeces and urine. Until recently, urine was heavily in demand for tanning leather and used in curing textiles, as evinced by memories of wearing an old tweed coat in the rain. Moraji Desai caused something of a stir during his tenure as India's Prime Minister in 1978 when he advocated drinking one's own urine to maintain your health. Urine therapy goes back hundreds, if not thousands, of years and retains celebrity adherents to this day. While faeces is used in various forms for farming, this is looked on as a dreadful trade due to the health hazards of working with it. Take a look at English excretory exclamations. Saying 'piss off' to somebody counts as swearing, but at the mildly sneering end of the scale. But if somebody says you're 'talking crap' or 'a piece of shit', offence is meant. Perhaps rightly, excrement remains a taboo subject, because of the harm it can cause. In his book *London: The Biography*, Peter Ackroyd makes play with sewage and prisoners.[8] Like prisoners, sewers are not meant to be talked about in polite society, and while sewage may be strictly treated, rehabilitation is out of the question. Maybe it is time to question some of our received wisdom.

As we will see, there is an elegant circularity here when it comes to environmental and health concerns. A former partner of one of London's more happily unreconstructed and traditional law firms pointed out to me that what I considered to be environmental legislation was, in the eyes of many in his profession, a public health matter. What used to be seen as a matter of aesthetics became a matter of life or death. Today environmental law in developed countries and its allied concerns are of a somewhat discretionary nature, about how much people are prepared to spend in the hope of attaining various desirable outcomes. It is fascinating to see how this comes back to public health, given that sustainability may well indeed be a matter of economic or, indeed, human survival in the face of various threats, rather than issues of convenience or aesthetic pleasure. Sanitation used to be thought as concerning matters foul rather than fatal, but through an accidental confluence of effluents, the miasma that affected public policy had to be broken through.

A primer on sewerage and sewage treatment

Sewerage is the process whereby wastewaters are taken from where they are discharged to where they will be treated or disposed. Originally wastewater (also known as foul sewerage or sanitary sewerage) and surface water run-off (rain or storm sewerage) systems were often combined, but more intense rainfall – plus increased run-off where urban areas have been concreted over and cannot absorb rainwater and have a minimal absorptive capacity – has made the risk of foul sewer flooding unacceptably high in many urban areas. New sewerage systems typically have separate storm and foul sewerage networks where heavy rainfall and the lack of surface percolation raises the potential for sewer flooding.

Wastewater is taken to a sewage treatment works (also known as wastewater treatment works or, as we shall see in the next chapter, water recovery works). Here it may be mixed with trade effluents, which are dilute wastewater discharged by industry into the sewerage network. Sewage is treated to a number of levels, usually in distinct phases, each progressively reducing its environmental and health impact. Primary treatment involves getting the solids, the brown stuff, out of the flow; secondary treatment lowers the environmental impact of the effluents; tertiary and advanced treatment is about purification. There is a pretty close relationship between how much of the solids are removed and the lowering of the effluent stream's biochemical oxygen demand. The former is about aesthetics and the latter about the potential for effluents to deplete a river of its oxygen. When it comes to bacteria, the impact of more advanced treatment is dramatic.

How wastewater treatment deals with sewage's pollution loading

After treatment, sewerage water can be discharged into rivers or the sea or given further treatment for recovery. Water discharged into the natural environment ought to meet various standards set, for example, by the World Health Organization. This water helps to maintain river flow and where water is abstracted downstream and via the 'magic mile' it can be indirectly reused.

Level of treatment	Process involved/Faecal coliforms per 100 ml of effluent	BOD removal/ Solids removal (range)
None and preliminary	This removes solids flushed down lavatories, such as loo paper, condoms, tampons and nappies, lowering the aesthetic impact of the sludge but not its environmental impact.	0–5%
		2% (0–5%)
	1,000,000,000	
Primary	Physical treatment, the effluent is placed in a settlement tank, so that solids are left behind and the liquid effluent is then discharged.	2–35%
	10,000,000	30% (10–40%)
Secondary	Biological treatment using bacteria, where the effluent trickles through inert materials such as granulated carbon, gravel or more recently, moulded plastic, so that it comes into contact with microorganisms, which oxidise and clarify the effluent.	75–90%
	1,000,000	90–95%
Tertiary and advanced	This usually means chemical treatment (coagulation / flocculation / ion exchange) for the removal of nutrients such as nitrogen and phosphorous. Advanced treatment involves filtration and / or disinfection with ultra violet light or ozone before its final discharge.	95–98%
		99–100%
	Less than 2	

Sewage sludge or biosolids can be processed for recycling on agricultural land or, after further treatment, used as compost sold for horticulture and domestic gardens. Biosolids has a better ring to it than refined human excreta. These processes include digestion, where sludge is heated to 40°C to reduce the number of bacteria and pathogens, and pasteurising – heating the sludge to 60°C for several days so that all pathogens and bacteria are removed, making it satisfactory for a wide range of agricultural applications. This can be achieved by anaerobic digestion or composting. Anaerobic digestion generates methane which can be used to power the treatment process.

Sewage is harmful to human health and is generated in quantities that can make urban life unsustainable without appropriate handling and treatment. Sewage treatment techniques have been developed that can effectively deal with the public health and environmental threats it carries. Their application remains another question.

A brief history of sewers and sewerage

Even when nothing was done about sewage, people could take mortal offence when it came too close. My wife is a Gunn, descended from a clan of Viking origin based in Scotland's far north-east. Their motto, 'Aux Pax, Aux Bellum' means 'either peace or war', and their reputation for war-mongering extends back a millennium. In 1562, Alistair Gunn, son of the clan chief, declined to yield to James Stewart, the Earl of Moray, on Aberdeen High Street. Stewart, who was also the son-in-law of the Chief of Clan Keith (the Gunns' bitterest rivals), had to step into the sewage steeped part of the street: a fatal breach of etiquette. Alistair Gunn was pursued, caught and taken to Inverness, where he was executed after a mock trial.

Given that the mysteries of the water cycle were progressively unlocked between the sixteenth and eighteen centuries and that empirical science flourished in Britain from the seventeenth century, aided by the microscope and the flourishing of Enlightenment thinking in the eighteenth, it is perhaps symptomatic of wastewater's status that its role in spreading disease was not understood until the end of the nineteenth century. Until pretty recently, disease was regarded as an internal malaise, something that arose from the mismatch of humours, with cures ranging from purges to bleeding. Such an approach meant that for centuries causality was overlooked.

Ancient sewers

Today, Scara Brae in the Orkney Island's Bay of Skaill looks like isolation personified, eight stone houses in dunes and grassland on the edge of the sea. Five thousand two hundred years ago, that was not the case; it offered security and comfort to a group of people for over 600 years and a degree of sanitation rarely exceeded until recently, with each house having a conduit for the removal of wastes. The fact that such a small settlement went to these lengths suggests a period of stability and prosperity.

The Indus Valley culture is fascinating both because of how little we know about it and how advanced they were. What is notable about the cities of the Indus Valley is that sewerage was universal, the health implications of which we have only recently fully understood. The Indus Valley houses were connected to open channels running down the centre of the streets. Wastewater would be passed through pipes into covered sumps where solids would settle out and could be removed.[9] The systems were for both rain and foul sewerage. It served them well for centuries, although it looks as if the Indus Valley culture was in severe decline – due to a combination of floods and water shortages – when the Indo-Europeans descended upon the valley in 1700 BC.

In many ancient civilisations, planned sanitation appears to have been widely adopted, brought about by societies run by command and control. Water pollution was considered a sin in ancient Persia. Rainwater and urban water run-off was collected for reuse and directed into wells to recharge the groundwater. Knossos, a palace-city in Minoan Crete (2800–1100 BC), had two conduits, one for rainwater and the other for foul water.[10] The Babylonian cities of Ur and Babylon had vaulted sewers for household wastes and gutters and rains to deal with surface run-off.[11] Urban hygiene in many ancient cities owed more to taboos about moral decay rather than concerns about actual physical uncleanness.[12] In that sense, taboos succeeded in imposing what we would see as common sense in implementing urban policy. Urban drainage has also been seen in Assyrian, Etruscan, Chinese, Greek and Hittite cities.

The Roman Empire built upon this past experience when it came to water and sewerage networks. Its contribution lay in the scale and complexity of these services. Steven Solomon suggests that 700–900 litres per capita of usually potable water flowed to each Roman every day, enough to cleanse the city of a million of its physical malaises.[13] The common people might only have obtained 10% of Rome's water supplies, but this was enough for their needs and more than any other city could offer and more than many latter-day patricians deign to offer their subjects today. Roman engineering was also distinguished firstly by designing roads with storm water run-off in mind, then by the development of local sewers leading to larger sewers and finally, through the linkage of water supplies to sewerage, especially when dealing with excess supplies. Between 410 and 600, Rome's aqueducts and sewers had collapsed under the weight of invasions, and the city followed suit. At the time of the reunification of the Papacy in 1417, just 40,000 Romans remained. Over the next two centuries a series of 'Water Popes' revived the city's aqueducts and fountains and the city's fortunes in turn revived.

After Scara Brae, there was no sewerage in Britain for nearly 2,700 years. The Romans pioneered large-scale sewers, most notably in York and Colchester, although there does not appear to be hard evidence of their being any laid in London. Once the Romans left in 407, these cities were soon abandoned and the sewers with them. Peter Ackroyd sees matters getting pretty grim soon afterwards with excrement literally everywhere in Saxon and Viking London.[14] Across Europe, as the Roman Empire imploded, the sewer systems fell into disuse. They remained the apex of sewerage systems for over 1,400 years. In this case, ignorance brought little in the way of bliss.

Several early cities employed sewerage systems to an extent that has only recently been exceeded. While they may not have been exemplars of an inclusive civic society, their leaders did at least recognise the benefits of sanitation as an integral part of urban civilisation.

The medieval malaise

In medieval cities, sewers were part of drainage pathways, typically found down the middle of a road. Their purpose was to let rainwater and whatever lay on the road drain to the nearest river or stream. The first covered sewer in Paris[15] was built by Hugues Aubriot in 1370, taking sewage from the Right Bank and discharging it into the Seine near the Louvre, where its stink reached the highest levels of society. By 1636 there were 24 covered sewers in Paris[16] all of which were clogged. Acts passed in 1721, 1736 and 1755 made property owners liable for cleaning the sewers that ran beneath their buildings – which made matters worse as, despite various prohibitions, this was interpreted as allowing them to dump all household wastes into these systems. These underground sewers symbolised the underclass, unnoticed except for the fact that sewers and the lower orders were a constant threat. When cholera and civic unrest erupted in 1832 and 1849, all suffered.

Medieval London was probably worse off than Paris. Between the thirteenth and sixteenth centuries, London had three streets called Pissing Lane or Pissing Alley, along with a Dunghill Lane and Dunghill Stairs. Some still exist, like Gutter Lane, Staining Lane and Seething Lane, while EC4's Sherborne Lane was Shiteburn Lane 400 years ago. London's first public loo, or house of easement, opened in the thirteenth century, discharging straight into the river, while the better-off could pay a tax to defecate directly into the Thames.[17] By 1400, there were eight public loos serving a city heaving with 100,000 people – including one with two rows of 64 seats at the Thames-side end of Friar's

Lane, along with private privies for the more affluent. Holding Frankincense to the nose was popular because it could mask the smell of excreta and when the smell overpowered these herbs, people started complaining.

Two cases from the London Assize of Nuisance, a court overseen by the Mayor to administer the smooth running of the city, give a flavour of the time. In August 1314 the Assize noted: 'a gutter (gutera) running under certain of the houses was provided to receive the rainwater and other water draining from the houses, gutters and street, so that the flow might cleanse the privy (camera privata) on the Hithe, Alice Wade has made a wooden pipe (pipam ligneam) connecting the seat (sedile) of the privy in her solar with the gutter, which is frequently stopped up by the filth therefrom, and the neighbours under whose houses the gutter runs are greatly inconvenienced by the stench. Judgment that she remove the pipe *within 40 days etc.*'[18] While in 1347: 'John le Yonge complains that Henry le Yonge and John Conyng have a solar above his cellar in the par. of St. Mary de Abbechirche, and the pipe of their latrine is in the same cellar and overflows into it... it is adjudged that *within 40 days etc.* they remove the nuisance.'[19] This was ad hoc justice in a city of post hoc sanitation.

At King Edward III's urging, four scavengers were appointed in each ward in the mid-fourteenth century to keep the streets clear of filth. This had become an especial problem in 1349 when labour became scarce in the wake of the Black Death. In 1357, throwing waste into the Thames or other waterways was outlawed. In reality, progress was piecemeal and English sanitation regulations remained effectively unchanged between the appointment of Commissioners for Sewers in 1427 and the 1848 Public

Health Act. Since the Commissioners were left unpaid until 1622, the impact of the Act was always limited.[20]

In comparison with many cities and civilisations of the ancient world, medieval Europe can be characterised at a time and place when sewage was managed on the hoof at the best of times. This would have increasing consequences as cities grew in size and importance.

Liberty or death? The early modern world

An unsentimental analysis of medieval England suggests that by leaving serfdom, the town dweller gained a degree of personal freedom and relief from poverty. That was counterbalanced by illness and death due to pollution, a case of disease in place of despotism. The prospect of some relief from poverty exploded with the industrial revolution, as did the unsanitary nature of city life. Eventually the sheer impact of human crowding would change the manner in which water and sewerage were managed. We will never get the whole picture as to how many died due to waterborne diseases in the past. In the UK, censuses provided a snapshot every decade from 1801, but many infants died between each tally of the still living. Electoral rolls kept regular tabs on the population, which was splendid if you were male, of mature age and with a standing commensurate with the right to vote. Otherwise, you were a non-person in the eyes of the state, a state of affairs seen across the world to this day.

Changing habits and wealth don't have a great impact on sewage generation. What matters is what happens to it and where we live: the greater the proximity, the greater the concern. From the 1450s, human dung was carted away from London and used as fertiliser on the fields outside the city. This was at the best of times a disgusting job – well paid for the carter and free fertiliser for the farmer, but coming at a mighty cost to their health. Urban growth made carting away sewage increasingly impractical, given the ability of the land to hold so much of the stuff. This in time forced the development of sewage farms and sewage treatment. Problems in Britain were exacerbated by a new source of fertiliser since from 1850 at least 100,000 tonnes of high quality Peruvian Guano was imported into the UK every year. Guano is stable, dry and relatively free of pathogens. Demand for sewage as a fertiliser collapsed in its wake.[21]

Cesspits and privy vaults tended to leak, when not overflowing, and contaminated nearby wells. As water usage increased, sewage sumps became increasingly overloaded. The irony here was that this was often due to improved public hygiene, so that taking baths was outlawed in Boston and New York without a doctor's permission.[22] Technology had a role to play in pollution. John Harrington built the first modern flush lavatories in 1596, one for Queen Elizabeth, his Godmother, and one for himself. An improved version was developed in 1775 by Alexander Cummings and between 1778 and 1797 Joseph Bramah sold 6,000 improved lavatories. In 1861 Thomas Crapper developed a unit that delivered 'a certain flush with every pull' and gave US troops returning from the Great War in 1918 a new slang word.

The Thames itself stayed fairly clean until 1815–30. As cesspits were progressively overwhelmed by developments such as the flushing lavatory – along with increasing overcrowding and shoddy building – household drains were connected to the city's rain sewers. The flush water mixed with other waters discharging into the river, while backing up during high tides. The last salmon was caught in the Thames in 1833.

From 1834, the Office of the Registrar General in the UK produced hard data on mortality in a series of Weekly Returns. As campaigners suspected, the death dates in major cities were unsustainable, even if nobody knew just why they were so high. Edwin Chadwick observed that peace-time losses from urban squalor exceeded those of any war Britain had recently engaged in. People preferred their water to be clean and wholesome, but this was seen as a matter of convenience, rather than one of life or death. Concerted plans to clean up cities were often a case of emergency and expediency rather than actually changing matters.

More people using more water made their rivers increasingly unusable from the nineteenth century. This was seen as a matter of aesthetics at the time, but it would later be seen as a pressing public health concern.

Sewers and sewerage

In Europe and Asia, dry collection was traditionally favoured because sewage could be put to agricultural use, although there was less enthusiasm for this in North America. The first centralised sewerage system was developed in Hamburg in 1843 after the city had been hit by a fire the year before, providing the chance to build a cleaner city.[24] Eight more German cities were to enjoy sewerage schemes in the next 21 years. In its wake came schemes for London, Chicago and New York's Brooklyn and, from the 1850s, there was a broad shift in favour towards sewerage, especially where planned systems were developed. The revival of sewerage in the modern world can be perhaps traced back to the private wooden sewers which were constructed in Boston in the late seventeenth century, and in 1704 a private sewer 'beneficial to the common citizen' was built, with those connected paying for its construction.[25] Between 1708 and 1736, 654 sewer permits were issued and this ad hoc approach made Boston amongst 'the most dry and clean cities in the world'. These sewers were for rainwater, with cesspits and privy pools dealing with the hard stuff.

In time, sewerage would be seen as the more cost-effective option given the long-term costs of developing and maintaining a sewerage network compared with cleaning and emptying cesspits and latrines. Lowered costs, greater convenience and getting the sewage swiftly out of sight made sewered cities more attractive than the status quo. Anyway, there were no other viable approaches at the time, so sewerage became a compelling alternative to what was seen as unacceptable state of affairs.

There are two TLAs (triple letter acronyms) in sewers and sewerage; SSS or separate sewer systems and CSO, or combined sewer outflows. Aesthetics drove the original sewerage schemes. Separate sewers for storm (rain) and foul (sewage) water were proposed at the same time as combined overflow systems, but their applicability was limited by the building techniques of the time. Edwin

Chadwick and John Phillips proposed separate sewerage for London in 1849, but using the brick-lined systems commonplace then this system would have cost much more. In the long run, the development of vitrified clay pipes against brick-lined conduits has eroded the cost difference between the two systems. The first separate system to be developed in the USA was for Lenox, Massachusetts by George Waring Jr. in 1875, with a larger system serving Memphis, Tennessee in 1880. There was a tendency in the USA to use CSOs for densely built-up areas and SSS for areas where rainwater did not need to be moved underground. Between 1860 and 1900 the number of people served by sewerage in the USA grew from 1 million to 25 million.

Combined sewerage networks are still used in the USA, Germany, Japan and Switzerland, where the emphasis is on controlling the inflow of stormwater to prevent systems overloading. Since the Second World War there has been something of a shift back to more decentralised approaches in the USA, due to the emphasis on low-density housing developments that are best served by independent sewerage networks supplying a dedicated wastewater treatment works. Today, such systems serve 25% of the population and account for 37% of new developments. This reflects a wider debate about the financial and energy costs of getting sewage to a major sewage treatment works (for example, serving more than a million people) against the lower costs of treatment due to the economies of scale possible at a large facility.

Cholera and choleric mindsets

According to the *British Medical Journal* in 2007, sanitation is the 'most important medical advance in the past 150 years.'[26] That advance has been closely linked to the role modern sewerage has had in lowering the impact of cholera in major cities. Cholera was first noted in the Ganges in 1817. Being a disease spread by faeces contaminating drinking water, burgeoning trade links rapidly spread it to cities such as London, with cholera epidemics hitting London in 1831–32, 1848–49 and 1853–54. The latter two outbreaks in London claimed 25,175 lives. Seven cholera pandemics have swept from the Ganga and Indonesia between 1817 and 1961, with the latest pandemic still going strong after some 50 years.

At the time, there were two broad theories about the transmission of diseases such as cholera. The miasmic or anticontagionists believed that noxious gasses or miasmas from sewage gave off these deadly effusions and that sewage wastes needed to be removed within two to three days if miasmas were not to emerge. They also urged improved street cleaning and proper ventilation for houses. The contagionists believed the disease was spread by people and urged quarantining the sick. In the case of cholera, they were equally and catastrophically wrong. As a result drinking water was ignored, and initial attempts to prevent the transmission of the disease made matters worse, since this merely resulted in more sewage entering drinking water sources. Both theories fitted in with a mindset that blamed the poor for getting what they deserved, as they were seen as ignorant and dirty. Charles Rosenberg quotes from an editorial about a poor district in Manhattan in the mid-nineteenth century: 'Be the air pure from Heaven, their breath would Contaminate it and infest it with disease.'[27] But did these people want to live that way? That is best summed up by a letter sent by 54 East Enders to *The Times* in 1849:[28]

'THE EDITUR OF THE TIMES PAPER, Teusday, Juley 3, 1849
...We are Sur, as it may be, livin in a Wilderniss, so far as the rest of London knows anything of us, or as the rich and great people care about. We live in muck and filthe. We aint got no privies, no dust bins, no drains, no water splies, and no drain or suer in the hole place. The Suer Company in Greek Street, Soho Square, all great rich and powerful men, take no notice watsotnedever of our complaints. The Stenche of a Gully-hole is disgustin. We all of us suffur, and numbers are ill, and if the Colera comes Lord help us...'

Not having any votes, it was no surprise that their concerns went unnoticed, even when cholera did return. But others were concerned about the real cause of cholera and dealing with it. John Snow's first publication *On the Mode of Communication of Cholera* made little impact in 1849. His next was to have a greater effect; Snow's 'ghost map' shows deaths from cholera around Broad Street between 19 August and 30 September 1854. Snow simplified the street layout, highlighting the thirteen water pumps serving the area and representing each death by a black bar, in order to demonstrate how cholera was spreading – not by any miasma,[29] but in water contaminated by human waste. This way, he was able to show the compelling cluster of fatalities among people using a single free well pump in Broad Street, Soho, while other pumps in Soho attracted but a fraction of these fatalities. He managed to have the pump closed in 1854, but the prevalence of the miasmic theory prevented any further action being taken, despite being identified by Filippo Pacini in the same year and even though the London epidemic in 1866 only affected those areas which had yet to be connected to Bazalgette's sewerage scheme, further vindicating Snow's map over miasma. It would be another four decades before it was agreed that cholera was a waterborne disease. Thus at the 1874 international sanitary conference delegates voted unanimously to declare 'ambient air is the principal vehicle of the generative agent of cholera.'

Establishing how cholera was in fact spread proved elusive. While Louis Pasteur was able to demonstrate how bacterial contamination of food and drink came from microorganisms in 1864, he was unable to identify the cause of cholera during an epidemic the next year in Marseilles. Between 1878 and 1882 Robert Koch demonstrated how blood poisoning came from the septicaemia bacteria and identified the tuberculosis germ. Using these techniques, Koch was able to describe the cholera bacillus to the Berlin Imperial Cholera Commission and the 'Conference for the discussion of the cholera question' in 1884 after examining cholera cases in India and Egypt over the previous year.[30] His belief that cholera was transmitted in drinking water was rejected by French and British delegations.

After the cholera outbreak in Hamburg in 1892 Koch was asked to investigate. In 1893 he isolated the bacillus in the Elbe and demonstrated that while it remained in

Hamburg's unfiltered water, it was absent in filtered water provided to neighbouring Altona. Hamburg had pioneered sewerage 50 years earlier, but there is a difference between what you take out and what you put in.

From then on, a gradual consensus about the real way in which cholera is spread emerged. In the end, history did not end with a bang, but a harrumph. It was one of those debates that had no definitive beginning, let alone an end. Perhaps the realisation that cholera was transferred in drinking water was conveyed by a process of miasma.

Sewerage emerged as a way of making cities more pleasant places to live in. While they were part of a drive to improve public health, the connection between sewerage and disease prevention was not generally appreciated at this time.

Bring on Bazalgette

Edwin Chadwick's Report on the Sanitary Condition of the Labouring Population of 1842 linked the need for adequate water and sanitation to combat ill-health in urban Britain. He chaired a central board of health from 1848. Chadwick believed that every household needed its own tap and sewerage connection. 160 years on, this remains a laudable recommendation.

Sometimes parliamentarians need a dose of reality before they are prepared to act.[31] Anybody living near the Thames in 1850 knew something had to be done, but Parliamentarians turned their noses up at these concerns until the Houses of Parliament were closed in June 1858 due to the overwhelming odour of raw sewage oozing down a putrid Thames. Only then was sewerage taken seriously in London. One hundred and thirty-seven schemes to clean up the Thames had been proposed by the time of the Great Stink, but all had been turned down.[32] Perhaps that was a blessing, as it allowed Joseph Bazalgette, who had been examining the proposals submitted to the Sewers Commission since 1850 and who was promoted to Chief Engineer of the Metropolitan Board of Works in 1856, to take action. 'Parliament was all but compelled to legislate upon the great London nuisance by the force of sheer stench,' noted *The Times* on 17th June 2008, in celebrating the day's 150th anniversary. The enabling legislation for a £3 million scheme was passed in just 18 days.[33] Bazalgette's scheme involved connecting all of London's sewers into two sewer mains running on the north and south embankments of the river. In fact, it would cost £20 million or £1.85 billion in today's money – the work lasting sixteen years, building 82 miles of sewer mains lined by 300 million bricks and forming the embankments that mark the centre of the city today. Intriguingly, the Weekly Returns noted no rise in Cholera during the Great Stink as the miasmists has assumed.[34]

Cleaning up Britain

Life expectancy in England and Wales rose from 40 years in 1847 to 50 from 1880 and onwards to 62 by 1912. Despite Bazalgette's work in London, the infant mortality remained mired between 138 and 163 deaths per 1,000 live births between 1841 and 1900, yet had fallen to 100 by 1912. This was due to the great expansion in sewerage across the whole of urban Britain in those latter decades as average annual spending on sewerage rose from £19 million between 1881–95 to £43 million between 1895 and 1910. The relationship between the two may not be irrefutably causal, but it is far more than coincidental.[35] The period of stagnation in infant mortality took place when incomes doubled, demonstrating it is not wealth that always matters, but how it is applied. In contrast, average incomes rose by a mere 6% between 1900 and 1912, at a time when infant mortality was falling at its fastest.

The long-term impact of Bazalgette and his successors across Britain continues to provide some of the most compelling evidence about how systematic sewerage implementation can improve life expectancy and lower infant mortality.

Towards modern wastewater treatment

The flip side of sewerage is what happens downstream. One point of view is that (despite or because of the 'magic mile's' best endeavours) water taken downstream can be treated irrespective of what has been added to it. In Britain, Bazalgette's scheme simply shifted the problem elsewhere. Human exposure to the sewage was as catastrophic as ever. In 1878, the Princess Alice, a paddle steamer went on a pleasure cruise and sunk in an accident near the Beckton sewage works; many of the 640 deaths were from poisoning rather than drowning. As a result, sewage removal and treatment rather than downriver diversion started to be taken more seriously. Direct sewage discharges into the Thames ended in 1887, with the post treatment sludge being dumped at sea.

Sewage is costlier to transport and treat than water and, since its benefits are not as immediately obvious to the consumer as for improved water; there is a constant pressure to perform the minimum treatment required to comply with the law rather than to minimise the impact of sewage on the environment. Environmental costs were not taken into account and public health was an incomplete science. So, municipalities tended to regard wastewater treatment as an unnecessary extravagance while doctors and health boards argued for them on public health grounds. An engineer in 1903 opined that 'it is often more equitable to all concerned for an upper riparian city to discharge its sewage into a stream and a lower riparian city to filter the water of the same stream for a domestic supply, than for the former city to be forced to put in a wastewater treatment works.'[36] Such an argument prevailed in the USA for some time. In 1892, there were just 27 cities in the USA with wastewater treatment works, 26 treating wastewater from separate sewer systems and one from a CSO.[37] Six used what was called chemical precipitation (secondary treatment) and 21 using sewage farms. By 1905, 95% of America's urban wastewater was untreated, and even in 1924, 88% of wastewater in cities of over 100,000 was untreated. It was not until 1960 that more than half of Americans had their sewage treated.

In 1858, the Sewage Commission noted that 'perfect purity of the water of a large tidal river, such as the Thames, passing through a densely populated district before reaching the metropolis...could not be expected even if the entire sewage of the metropolis itself were removed.'[38] But things can certainly be improved. Apart from the increased understanding about other diseases, such as typhoid being transmitted downstream in untreated wastewater in the 1890s, the development of techniques such as activated sludge management made larger scale sewage treatment more affordable.

The patricians who ran public health saw city dwellers in a poor light as they believed that 'the lower classes of people cannot be allowed to have anything to do with their own sanitary arrangements: everything must be managed for them.'[39] That attitude prevails, and the blithe assumption that people will put up with anything when those who act on their behalf have more important things to do actuate water and sanitation policy to this day. The World Health Organization has pointed out that while early sanitary engineers can be excused for ignorance about pathogens in foul water, the same cannot be said for contemporary engineers.

Here historic legacies lie. There was a conflict between industrialists who were happy to pay for water to fuel their factory processes but not to improve the lot of their factory workers. When the latter started gaining the vote, they became a constituency and universal water and sanitation entered into the consciousness of the Commons as it did at the start of the twentieth century. The development of Britain's economy, political franchise and public health has lost none of its resonance across the past century. Perhaps too many countries are afraid of broader representation due to its implications about sanitary expectations.

Public concern has become a major driver towards proper sewerage and sewage treatment and in much of the world these concerns still need to be properly heard.

From folklore to science

As John Sidwick observed, Britain's 'Royal Commission (on the treatment and disposal of sewage) marked the transition from folklore to a scientific approach to sewage treatment', meeting in various forms over 51 years.[40] The ninth Royal Commission was appointed in 1898, following on from eight such Royal Commissions established between 1857 and 1882.[41] The earlier Commissions reflected the then concerns about the chemical impact of sewage and concluded that land based disposal was appropriate. Since then, the role of bacteria in sewage started to be better understood, meaning that while land disposal remained an option, treating the sewage itself prior to disposal became a greater priority.[42] Where the soils are clayey, filtration was seen as the better option – as was often the case near London.

The Royal Commission's legacy lies in their recommendation that all that mattered was the quality of the effluent discharged, rather than whatever treatment process was used.[43] The 20:30 standard, that effluents should not have a Biological Oxygen Demand of more than 20 milligrams per litre of water and suspended solids not more than 30 mg/l, has been adopted worldwide and has been a worldwide force for the good. This is not an idealistic standard; it is a realistic one, concentrating on water quality rather than seeking aspirational standards of odour control and by-product elimination. This shifted the emphasis towards value for money and the 'magic mile'. It may not have been perfect, but it was achievable. In practice this meant that septic tanks fell out of favour except for small, isolated communities, while sedimentation, the practice of allowing sewage solids to settle out (physical or primary treatment), became the standard procedure before treating the liquid effluents.

The adoption of sewage treatment marks the completion of a first phase in humanity seeking to remedy the effects of urbanisation on the water cycle in terms of replicating the cycle's ecosystem services. It marked the realisation that sewage needed to be properly treated and disposed of. The extent of that treatment and disposal remains a contentious area to this day.

Treatment – technologies and techniques

Originally sewerage was about getting the sewage out of towns and cities rather than treating it. Getting the sewage effluents away from towns and cities is one thing, making it safe and ideally reusing it is another. The idea of applying raw sewage to land at sewage farms was that in time it would naturally digest, breaking down into useful nutrients. There were approximately 50 sewage farms in England by 1875, where sewage was formally delivered for land application. Other sewage was dumped into the sea, a practice that was finally ended in the UK in 1998.

Sewage treatment, rather than land application, emerged in the 1880s, originally using sand filtration and chemical precipitation. In 1882 Dr Angus Smith was the first to investigate how blowing air into sewage might improve its treatment. Until aeration was combined with filter beds (which unknown to them had their own populations of bacteria) it was not a great success in treating sludges.

One of the pioneering techniques was anaerobic digestion which works by accelerating the biological processes which digest sewage sludge.[44] In 1913, Sir Gilbert John Fowler and a Mrs Mumford (a rare being at the time, a female graduate engineering student) experimented with a suspended-culture aeration system using a specific bacterium called the 'M-7'. The process advanced a year later when Messers Arden and Lockett two of Sir Gilbert's students at his Manchester lab looked at saving the humus solids used, partly in order to use the sludge as a safe source of nutrients. Ten activated sludge treatment facilities were installed in England between 1914 and 1921 with the first being built in the USA in 1916.

Development was then curtailed for some time because of disputes arising from Patents granted to the engineers Jones & Attwood Ltd between 1913 and 1914, who were involved in several of the English projects and an American Parent granted to Leslie Frank in 1915. Jones & Attwood's Activated Sludge Ltd took various companies and municipalities in the USA to court in the 1930s. In the end, 150 of the 203 plants agreed to pay Activated Sludge a royalty of 25 cents per capita. Others fought them, developed alternative variants of the process (such as the trickling filter) or waited for the patents to expire.

In Germany, sewage fields were used from 1871 (Danzig), followed by screening and settling in Frankfurt in 1887, chemical precipitation in Leipzig in 1897 and the Imhoff Tank in Essen in 1908, which separates sewage sludge from wastewater for anaerobic digestion.

There was a constant flow of ideas and visits between Europe and the USA. The first use of a sand filter was at Medford, Mass. in 1887 and this was followed by a trickling filter (an Imhoff Tank) in Madison, Wisconsin in 1909, followed by the use of chlorination for effluent disinfection in 1914 and activated sludge treatment in San Marcos, Texas in 1916. It is remarkable how little these techniques have evolved, which shows how much was got right at the outset. However, dealing with the sludges themselves remains a perennial problem.

Following the introduction of tertiary treatment in 1920, there have been no fundamental changes in sewage treatment although a great deal of effort has gone into advancing treatment techniques, minimising effluent discharges and developing sludge and water recovery systems.[45] How much needs to be changed? When something works, that is what matters. Even so, the need for innovation, especially in reducing the energy needs of sewage treatment remains urgent, but not as urgent as the need for innovation in attitudes.

Sewage treatment has been developed as a series of pragmatic approaches towards various needs such as solids and biochemical loading. Many of the techniques used today have been in place for more than 90 years, demonstrating the robustness of the original approaches.

A disease that still cannot speak its name

The problem of cholera is still with us. In May 2011, at least 250 people died when floods contaminated wells serving Yaounde, the capital of Cameroon. Just 65,000 of the city's two million inhabitants are connected to the public water supply and in the unserved areas there is no effective sewerage or sewage management.[46] The deaths in Cameroon are depressingly unremarkable. Every town and city without adequate sanitation facilities have the potential to collide with unsafe water, and this can be replicated over and over again. That is why hundreds of thousands die every year from a disease which has been all but eradicated from developed economies. In fact, it ought to be eradicable everywhere, since it is as eminently treatable as it is vile.

It is difficult to escape the politics of expediency, whichever century you live in. Today, tourists are money, so cholera outbreaks are often under-reported in case they deter the patronage of the wealthy. It is unfair to blame the UN's agencies for such misinformation, since they are carrying out so much unheralded work where politicians fear to tread. But they can only report about what they are asked to cover, rather than all the other stories. Rather than monitoring all sources of information first hand they can only appraise what is made available to them.

So where are we today?

The development of sewerage and sewage treatment in developed economies has been a long and complex one. In the end, it was spurred by undeniable public health arguments which were underpinned by the realisation that these were also essential if the integrity of the water cycle was to be sustained. I developed this table to show the contrast between the Millennium Development Goals' official estimates for safe sanitation connections in 2008 and the more concrete realities of those who are actually connected to a sewerage network and have their sewage treated to at least secondary standard. It is based on data I have collated over thirteen editions of the *Pinsent Masons Water Yearbook*, sewerage data from the country surveys collated by the UN's Joint Monitoring Project and GWI's Global Water Markets 2011.[47] Some countries have been excluded because of their size or where, as in Somalia, reliable data gathering was not possible. Even so, it covers 99.5% of urban dwellers.

Urban sewerage and secondary wastewater treatment in 2008

Region (million people)	Urban population	Connected to sewerage		Secondary WW treatment	
Western Europe	308	294	95%	283	92%
Eastern Europe & Central Asia	306	251	82%	101	33%
Middle East & North Africa	213	129	60%	70	33%
Sub-Saharan Africa	294	48	16%	9	3%
Asia & The Pacific	1,034	474	46%	295	28%
South Asia	468	136	29%	36	8%
North America	279	265	95%	221	79%
Latin America	448	291	65%	108	24%
Global total	3,350	1,888	56%	1,121	33%

It does not appear that such a table has been developed before. Given that it shows that nearly half the world is not connected to sewerage and that just one third have their sewage adequately treated, this is no great surprise. The JMP estimated 76% of urban dwellers were connected to safe sanitation including 57% in Southern Asia and 44% in Sub-Saharan Africa.

When you examine the country-by-country data, the contrast between, for example Chile's 92% household sewerage and 72% treatment and the 47% connection rate and 10% sewage treatment rate for people in Argentina is both a matter of political as well as economic development. As we will see later on, when water and sanitation are proxies for other people's ideology, it is humanity that misses out. Politics is a continuing theme here, as, for all their failings, the capitalist countries of Europe and North America had moved on from sanitation to sewage treatment, while Steven Solomon notes that the people's utopias of the Soviet Empire and China (Czechoslovakia excepted, due to its pre-war legacy) lacked effective sewage treatment.[48] I suspect he is a bit harsh, since there is quite a bit of kit out there, but it is clear that much of the infrastructure that was in place in the former Soviet Union didn't work by the 1990s. Even nineteen years after the fall of the Berlin Wall, 33% of urban sewage in Central & Eastern Europe is properly treated against 92% in Western Europe.

Coverage matters, since the real benefits of sanitation can only be achieved through comprehensive management.[49] The benefits accrued from 90% sanitation coverage can be negated if 10% of households still have to practice open defecation or open drain discharge. This is one of the effective points about universal affordability. In informal settlements, the onus needs to be on community-led initiatives with rewards towards effective realisation at the neighbourhood level. This highlights the danger of the partial approach adopted by the MDGs. Countries may be content to scrape away at the numbers of the unserved without actually delivering tangible benefits on a community by community level and the target mentality being produced by exercises such as the MDGs create their own difficulties. WaterAid refers to this as 'latrinisation', the construction of household latrines to satisfy development goals, rather than how they are meant to deliver sanitation services.[50] Latrines remain a viable approach as long as sewage storage and removal does the job. This requires appropriate technology along with compact and inexpensive emptying systems that can reach latrines in densely populated areas.

The more water a household has access to, the greater its sanitation options. There is little point in flush lavatories if there is not enough water to flush the waste safely away. Households 'served' by a communal standpipe get by with 15–30 litres per person per day, compared with 25–70 for yard taps and intermittent household connections and 90–180 for in-house connections with a full supply. Technology pushes sanitation the other way as well, with modern dual flush lavatories using 20–40 litres of water per person each day, compared to traditional full flush models, which flush 30–60 litres away every day.

Household sewerage is the ideal, or perhaps an ideal. What matters first is truly safe sanitation, which is why properly maintained communal blocks (separate entrances for men and women, hand washing facilities and all excreta either going into fully lined and maintained septic tanks or plumbed into the sewerage network) can make such a difference. Sewerage here can be a grid or lines of sewered blocks, which in the future may become the basis for a household sewerage network. These sewers could either connect to a local facility or go to a larger treatment works.

Wandering past the pricier hotels in Leh each morning, where the manager would wear a blazer and the evening's bottled beer was chilled, you'd see the dhobi wallahs flailing towels and bedsheets on boulders in a nearby stream. Inside were flush loos plumbed into septic tanks that retained the solids and allowed contaminated wastewater to flow into the streams. Without a sewer network, let alone water or sewage treatment, the river which once fed the valley had become a waste conduit. People would wait for the tankers to bring their drinking water.

The dry compost loo and bring-your-own sleeping bag routine at the Oriental in Changspa Village near Leh was more austere but better fitted the infrastructure's capabilities. Perhaps we need to look to leapfrog technologies and techniques. Why not a 'western-style' loo unit, which has a valve which would only open when you sit upon it, so the already low smell from the dry compost unit below is contained and there is nothing to see? As with septic tanks, self-discipline is needed when it comes to what gets put away. In contrast, while post treatment sludge today may tick various boxes when it comes to pathogens and heavy metals, its smell remains pretty grim, no matter how it is applied. At home, on certain weeks, fields near us become rank through this urban release.

Compared with water provision, access to sewerage and sewage treatment remains far less prevalent, especially in developing economies, yet this 'poor cousin' ought to be treated as an area of equal concern.

The politics of sewerage and sewage treatment

Can sewerage and sewage treatment only be delivered by the public sector using subsidies? That is the belief of PSIRU, a trade union sponsored policy group who wrote a quite useful study in 2008.[51] Unfortunately, it overlooked the realities, such as the fact that Britain's great sewerage expansion came about through private sector funding of municipal projects. Indeed in northern Europe, the slowest development of sewerage and sewage treatment has taken place in Northern Ireland and Eire, where water is 'free' (paid through the rates and Government subsidies), while full cost recovery is the norm in Scandinavia, England and Wales, Germany and the Netherlands, all of which enjoy well-developed networks. The challenge, as ever, is to seek new funding without alienating support through increased taxation. Back then, bonds were the answer. The long-redeemed certificates mainly interest collectors of the recherché today, but their legacy is life enhancing.

Public investment matters, indeed it is essential, as often is private investment. What matters more is public will. Sewerage clearly will always be a greater challenge than for water, due to the higher costs and the more indirect nature of the benefits, but the simplistic relationship does not always hold. Subsidies have been a feature of sewerage extension in Japan, South Korea and Taiwan where the connection rate has been appreciably lower than in Singapore, where full cost recovery has been the norm despite their development trajectories having much in common. I developed this table to show how each of these country's urban water, sewerage and wastewater treatment infrastructure has developed in the past 47 years, in relation to their funding (full cost recovery or government subsidies) and economic development.

Water and sewerage infrastructure[52] and GNI per capita (atlas method, current US$)

Year	Singapore	South Korea	Malaysia	Japan	UK
	FCR	Subsidy	Subsidy	Subsidy	FCR
1962	450	110	300	610	– [1]
1970	960 [1]	270	400	1,860 [1]	2,230 [2]
1980	4,910 [1]	1,810	1,830 [1]	10,510 [1]	8,510 [2]
1990	12,050 [2]	6,000 [1]	2,390 [1]	27,160 [1]	16,600 [2]
2000	23,350 [3]	9,910 [1]	3,450 [1]	34,620 [1]	25,910 [3]
2009	37,220 [3]	19,830 [2]	7,350 [1]	38,080 [3]	41,370 [3]

Key
[1] At least 90% urban households with water connection
[2] At least 90% urban households with piped sewerage
[3] At least 90% urban secondary sewage treatment

Unfortunately Taiwan cannot be included as it does not officially exist. Different levels of urbanisation need to be considered as Singapore was 100% urbanised in 2008, compared with 89% for the UK, 81% for South Korea, 70% in Malaysia and 67% in Japan and consistent disaggregated urban – rural economic data is not available. Even so, it is evident that in Gross National Income terms, Singapore's economic development has taken place at a time when its water and wastewater infrastructure was appreciably more extensive than in Malaysia, where the water and wastewater infrastructure have developed more slowly. Japan and South Korea also show similar trajectories of economic and water infrastructure development with Malaysia and Singapore respectively. Since 1990 its water and sewerage infrastructure and since 2000 for sewage treatment have been fully developed and since 2000 Singapore's GNI has been comparable with OECD member countries such as the United Kingdom, which have a fully developed water and wastewater management system.

This is not just an academic debate, since it involves fundamental issues about how we want sewage to be managed. For example, by developing piped sewerage networks connected to suitable wastewater treatment, poorer areas can rapidly be connected to these networks via a grid of properly managed community facilities.

Indifference to cost recovery and service extension in developing economies stems in part from the urban middle classes who 'cocoon themselves from the effects of poor sanitation, either by availing themselves of the minimal networked sewerage that does exist or by moving into self-contained colonies with independent sanitation provision.'[53]

When nothing is done, things happen

Sanitation is the water sector's 'poor cousin' which, when you consider the state the water sector is in, is not good news. The greatest challenge facing sewage treatment and sanitation is one of perception.[54] The connection between access to safe water and well-being is simple and straightforward, but without education the case for sanitation spending is more convoluted. Official targets and achievements highlight the semi-detached relationship between policymakers and human realities. Where in slums is the space for universal household sanitation? Well-built and managed communal blocks offer at least an intermediate answer. These also use full cost recovery in a simple and direct way via 'sweat equity' (offering your labour for free to cut construction costs) and by affordable user fees. In Africa, studies have found that up to 70% of urban sanitation facilities are provided by small-scale entrepreneurs, typically where no alternatives exist.[55] If nothing else exists, entrepreneurs will seek to make good the gap, even where there is the pretence that the formal private sector has no role to play here. If municipalities or governments have no interest in providing the service, either nature or humanity will have to deal with it. This is particularly noticeable when private sewage collectors convey their wastes to municipal facilities as in Dar es Salaam. This is an example of the providers' good, where there is no economic incentive for governments or communities to carry out a service, but if it makes sense to an entrepreneur, then there is a niche that can be exploited. Here, antipathy towards commercialisation, full cost recovery and using the private sector is therefore creating even more opportunities for private entrepreneurs.

One of the challenges here is to turn this service to the public good.[56] Can these entrepreneurs be encouraged to channel their activities another way? There has to be scope for bringing entrepreneurs into a formal system. Informal service providers such as these are in essence apolitical, since they do a job that has to be done since the opportunity exists. If it did not, they would not be doing it. But who is regulating the disposal of the collected sewage? When you move from collection to treatment and disposal the capital costs escalate and the needs for some economy of scale kick in.

Sewage does not remain in urban areas forever, as people will take it away because in the absence of formal sewerage services informal providers step into the void. This demonstrates how failures of municipal oversight provoke responses, no matter how unsatisfactory they may appear to be.

Cleaning up in our safe European home

I remember being at a conference back in 1990 when the head of the then European Economic Community's Environmental Directorate spoke with real feeling about the need to see the cleaning up of Europe's environment not being a matter of aesthetics but of human rights. It had been traditional for politicians to nod and smile and then quietly ignore the environmental directives. But not any more, since countries that ignore environmental legislation can have European Union funding withheld and fined €110,000 per day until suitable remedial work is done. But reaching European standards is not cheap.

The 1991 Urban Waste Water Treatment Directive is the most expensive of the EU's earlier environmental directives to date, due to its rigorous requirement for providing wastewater treatment infrastructure for urban areas. It calls for all towns with more than 2,000 population equivalent (a term used to describe pollution loadings) to have their sewage discharges treated to at least secondary standard, with higher standards for 'sensitive waters'. The EU has estimated that compliance will have cost the original member states €152bn by 2010, with the accession states having up to 2015 to comply. In contrast, the 2000 Water Framework Directive is perhaps the least quantifiable piece of EU legislation because it leaves the stipulation of water quality improvement targets to the member states' discretion and proposes no deadline for achieving the 'ultimate aim' of good ecological quality waters throughout the Community other than its 2012–15 initial testing period. The Commission estimated that the investments required of industry under the proposal would amount to €2 billion, compared with independent estimates of up to €200 billion. Pouring sewage into the sea is out of the question, as the 1976 Bathing Water Directive set out minimum hygiene standards for Europe's beaches and the 2006 revision in effect demands full treatment from 2015. Other directives covering landfill and waste incineration, as well as the content of sewage sludge, all affect the way sewage sludge is disposed of. This matters, with 367 million urban dwellers living within the enlarged EU in theory churning out 156–313 million cubic metres of sewage sludge every year.

Two examples show how changes such as these affect the way sewage is collected and treated. Firstly, there is the transformation of sewage treatment in Austria and then the development of sewerage systems in Korea, Hungary and Portugal.

In Austria, Secondary treatment rose from 25% in 1980 to 60% in 1990 and fell back to 2% by 2006, while tertiary treatment rose from 3% in 1980 to 7% in 1990 and 83% by 2006. Normally developing national wastewater treatment infrastructure is something talked about and put in place over decades. Here and in the Netherlands, as the country prepared to raise its standards, wastewater treatment moved from the exception to the norm in a decade from 1980. Then with the Urban Wastewater Treatment Directive in 1991, an entirely new generation of treatment systems replaced their recent endeavours in another decade and a half. When laws change, everything changes.

Over roughly a quarter of a century, urban sewerage in South Korea, Hungary and Portugal has transformed from being the exception to the norm. The evolution towards democracy may be a common theme here, but this is not necessarily the case, since 53% of the Czech Republic's sewage was treated in 1995 against 18% for Hungary, its marginally more liberal neighbour at the time. Meanwhile China's sewage treatment infrastructure development since 1990 has not been matched by developments in its civic society. South Korea reflects the country's industrialisation and development, in part aided by the 1988 Seoul Olympiad and the transition from military to civilian rule, with full sewage treatment rising from from 8% in 1980, to 41% in 1995 to 82% by 2006. After the

'Carnation Revolution' of 1974, Portugal became a democracy in 1976 and joined the European Economic Community in 1986 with full sewage treatment rising from 12% in 1990 to 52% in 2008. Hungary left the Soviet Bloc in 1989 and became a democracy in 1990, joining the European Union in 2004 with 15% full sewage treatment in 1990 and 42% by 2006. European Union law and funding has driven sewage treatment's development. In Portugal's case, as one of Western Europe's poorer countries it qualified for some serious financial assistance, garnering €3.9 billion in Structural and Cohesion funding from 1993–2006 for environmental projects and a further €1.4 billion for water and wastewater projects in 2007–13. Hungary by contrast was an Accession State, part of the eastwards expansion of the European Union. Complying with European laws was made a stricter process for the candidate states than the incumbents, with a firm adherence to timetables which had been missed by years or even decades. As a sweetener, €5.5 billion in funding for environmental and nuclear safety projects for the new member states was provided between 1990 and 2006, including €583 million for water projects in Hungary for 2004–06 along with €1,959 million for water projects budgeted for 2007–13.[57]

Attitudes towards standards of sewerage and sewage treatment are often a question of political will, especially when it comes to meeting EU standards. In 1957, Caroline Wakefield died aged six after contracting polio from raw sewage discharged into the sea at Eastbourne.[58] Her death and the official indifference to it spurred the creation of what is now the Marine Conservation Society, which publishes a *Good Beach Guide* every year. Britain fought tooth and nail against the effective implementation of the 1976 Bathing Waters Directive as seen by a grim series of articles in the *New Scientist* at the time.[59] It was the old conflict of the 'indifferent engineer' versus public health officials and the former prevailed until July 1993 when the European Court of Justice ruled that Blackpool was indeed a bathing area not a resort that incidentally faced a sandy beach. Since then a heroic amount of work has been carried out, and indeed, after years of political and planning delays, Eastbourne today counts amongst Britain's cleanest by the *Good Beach Guide*.[60] There still can be an ad hoc dimension here. Nicholas Ridley, the Environment Secretary in 1989, was a keen fly fisherman and sought to ensure that river water quality was improved as part of the 1989 privatisation settlement in England and Wales.[61]

The EU has made a remarkable impact both on the development of sewage treatment in Europe and Europe's attitudes towards sewage treatment. The political and public response to this has generally been phlegmatic and pragmatic and even in the UK, where such legislation was once opposed, there is a belated realisation that sewage treatment and management can provide benefits as well as incurring costs.

Back to the Thames...or the Thames gets backed up

Cleaning up the Thames took decades and cost a lot of money, yet it still remains a work in progress today. In 1974 a single dead salmon was seen on the river, the first since 1833.[62] Improvements in sewage treatment and a reintroduction programme in 1979 made an impact and a six-pound specimen was caught in 1983. The idea of the Thames Salmon Rehabilitation Scheme was to showcase that the river was healthy again. A decade later saw hundreds swimming up the river but in recent years they have since become scarce. Low upstream flows and contamination from the overloaded sewer network are to blame. That is the difference between the Thames and the Seine, Rhine and the Tyne, where sewerage systems can cope with heavy rainfall and salmon numbers have been successfully revived.

Today Thames Water is proposing the Thames Tideway Sewer, a £3.6–4.3 billion project to construct a 22 Km long and 7.5 metre wide sewer to augment the Bazalgette mains sewer, which after some 150 years has been overwhelmed by new demand. Oddly enough, Thames Water has been trying to start the scheme for a decade, but Ofwat fought shy of the bill in 2004. That August, heavy rains caused an overflow of ten million tonnes of sewage into the river and a subsequent fish kill. In engineering terms, the rains were almost providential; the project is back on track and may start in 2012–14 with completion by 2021. From Bills to bills, politicians and public perception ('Fulham says NO to the super sewer!') continue to obstruct London's sewerage development.[63] In the meantime, instead of 3.2 million people in 1860, there are 7.6 million in London today and 39 million m^3 of foul sewerage ends up in the Thames each year because there is nowhere else for it to go. What comes in goes out. This neatly sums up the limitations of the CSO approach. When more water goes into this type of system, it cannot cope with the additional volume, and foul water as well as rainwater gets discharged untreated. As the *Financial Times* noted in 2009: 'Thames Water's corporate motto – "if customers had a choice, they would choose us" – tells you all you need to know about the anomalous nature of the water sector.'[64]

Conclusions – science and sewerage go hand in hand

The history of men is reflected in the history of sewers...The sewer is the conscience of the city. Everything there converges and confronts everything else.

Victor Hugo, *Les Misérables*

Sewerage in early civilisations was notably less exceptional than for most of human history. One of the earliest great civilisations, the Indus Valley culture of North West India seems to have offered universal household water and sewerage. Rome and York are excellent examples of later sewerage systems, unsurpassed for the next 1,500 years. The Roman Empire was indeed about bread and circuses, but also about ensuring basic public services got provided.

Since then, the absence of sewerage placed a great premium on households straddling the riverbanks and, better still, being built on bridges. Without proper sewerage, flush loos can compound matters, dispersing the sewage amongst others rather than leaving it in a manner easily handled by scavengers. The norm has been that cities 'deal' with sewage by occasional rainfall or floods sweeping the problem away. This is not a particularly effective or reliable method and effluent loading was probably as much an impediment to urban growth as securing water supplies.

The stand-off between rural despotism and urban illness became unendurable as cities grew beyond a point where laissez faire sewerage management could reasonably deal with the wastes generated and death rates rose unsustainably. Old attitudes have prevented public health being taken seriously for centuries, especially when it comes to sewerage and sewage treatment. The poor suffered from poor health because it was assumed that it was propagated amongst them and was a punishment for their morals rather than a reality that affected all. Confronting such attitudes was far harder than it was for improving drinking water, but a combination of bad odours and Cholera proved compelling, resulting in a series of initiatives to remove sewage from cities and then to treat it.

Every stage of the development of sewerage and sewage treatment has been met with scepticism. That attitude can be seen in the reluctance to deal with raw sewage discharged into Britain's bathing waters and the minimal progress being made towards sewage treatment for urban dwellers in South Asia and Sub-Saharan Africa.

The technologies developed between 1840 and 1920 form the backbone of sewerage and sewage treatment today. That should not be a concern because in most cases they work and are the most appropriate approaches. There is great scope for improving the efficiency of systems, developing new processes to deal with new waste streams such as pharmaceutical wastes and towards sewage recovery, but the fundamentals are sound.

Systemic change is needed, not for the engineering but for managers and politicians and this will remain the case until sewerage and sewage treatment are the norm, rather than the exception.

From here we need to consider wastewater as a potential resource, rather than a nuisance. This will require looking at replicating natural wastewater treatment as well as more advanced approaches. Reed beds are pretty efficient at treating wastewater and where sewage is discharged onto an estuary, reed beds and brackish water can do a lot of work. It may well be that concerns about lost resources and the values embedded in them will act as a fundamental driver for taking wastewater treatment and recovery more seriously.

---

[1] Gotaas H B (1956) *Composting; sanitary disposal and reclamation of organic wastes*, WHO, Geneva, Switzerland

[2] Drangert J O (1998) Fighting the urine blindness to provide more sanitation options, *Water SA*, 24, 2, 157–164

[3] Tchobanoglous G & Burton F L, eds. (1991) *Wastewater Engineering, Treatment Disposal Reuse*, Metcalf & Eddy / McGraw-Hill, New York, USA

[4] Lester J & Edge D in Harrison R M, ed. (2001) *Pollution: Causes, Effects and Control*, 4th Edition, The Royal Society of Chemistry, London, UK

[5] House of Lords (1991) *Municipal Waste Water Treatment Vol. 1: Tenth Report of the Select Committee on European Communities*, 1990–1991. HMSO Books, London, UK

[6] Mara D (2003) *Domestic wastewater treatment in developing countries*, Earthscan, London, UK

[7] CGER (1996) Use of Reclaimed Water and Sludge in Food Crop Production, Water Science and Technology Board, Commission on Geosciences, Environment and Resources, National Research Council, National Academy Press, Washington DC, USA

[8] Ackroyd P (2000) *London, The Biography*, Chatto & Windus, London, UK

[9] Webster C (1962) The sewers of Mohenjo-Daro, *Journal Water*

Pollution Control Federation, 34(2), 116–123

[10] Gray H F (1940) Sewerage in ancient and mediaeval times, *Sewage Works Journal*, 12, 939–946

[11] Jones D E Jr. (1967) Urban hydrology – a redirection, *Civil Engineering*, 37 (8), 58–62.
Maner A W (1966) Public works in ancient Mesopotamia, *Civil Engineering*, 36 (7), 50–51.

[12] Burian S J & Edwards F G (2002) Historical Perspectives of Urban Drainage, Global Solutions for Urban Drainage; 9th International Conference on Urban Drainage, 8–13, September 2002, Portland, Oregon, USA

[13] Solomon S (2010) *Water: The Epic Struggle for Wealth, Power and Civilization*, Harper Collins, New York, USA

[14] Ackroyd P (2000) *London, The Biography*, Chatto & Windus, London, UK

[15] Reid, D (1991) *Paris sewers and sewermen*, Harvard University Press, Cambridge, MA, USA

[16] Krupa F (1991) Paris: urban sanitation before the 20th century, www.op.net/~uarts/krupa/alltextparis.html

[17] BBC (2011) An Asbo in 14th Century Britain, http://www.bbc.co.uk/news/magazine-12847529 accessed 5th April 2011

[18] 'Misc. Roll DD: 1 Mar 1314 – 1 Dec 1318 (nos 201-251)', *London assize of nuisance 1301-1431: A calendar* (1973), pp. 41–54. URL: http://www.british-history.ac.uk/report.aspx?compid=35973 Date accessed: 05 April 2011

[19] 'Misc. Roll DD: 14 Dec 1347 – 29 Jan 1356 (nos 400-449)', *London assize of nuisance 1301-1431: A calendar* (1973), pp. 99–110. URL: http://www.british-history.ac.uk/report.aspx?compid=35977 Date accessed: 05 April 2011

[20] Sidwick J M (1977) *A brief history of sewage treatment*, Thunderbird Enterprises, Ltd., Middlesex, England

[21] Vizcarra C (2006) Guano, Credible Commitments, and State Finances in Nineteenth-Century Peru, *Journal of Economic History*, September 2006

[22] Armstrong E L ed. (1976) *History of Public Works in the United States, 1776-1976*, American Public Works Association, Chicago, USA

[23] Burian S et al (2000) Urban Wastewater Management in the United States: Past, Present, and Future, *Journal of Urban Technology*, 7, 2 33–62

[24] Metcalf L and Eddy H P (1928) *American sewerage practice, volume I: design of sewers*, McGraw-Hill Book Company, Inc., New York, USA

[25] Armstrong E L ed. (1976) *History of public works in the United States: 1776-1976*, American Public Works Association Chicago,

Illinois, USA

[26] *British Medical Journal* (2007) Medical Milestones Supplement, *Brit Med J* 344 Suppl 1.

[27] Rosenberg C E (1962) *The Cholera Years: The United States in 1832, 1849 and 1866*, University of Chicago Press, Chicago, USA

[28] http://historyday.coldray.com/works-cited/#54citizens accessed 3rd June 2011

[29] Howard-Jones N (1984) Robert Koch and the cholera vibrio: a centenary, *British Medical Journal* 288, 379–381

[30] Koch R (1884) An address on cholera and its bacillus, Delivered before the Imperial Board of Health at Berlin, BMJ, 30th August 1884, 2, p403–407

[31] Halliday S (1999) *The Great Stink of London: Sir Joseph Bazalgette and the cleansing of the Victorian metropolis*, Sutton Publishing Limited, London, UK

[32] Simon P (2008) The big stench that saved London. 150 years ago, Victorian genius solved a stinker of a problem. Now the work needs to be finished. *The Times*, 17th June 2008.

[33] http://historyday.coldray.com/bazalgettes-actions/ accessed 3rd June 2011

[34] Shapin S (2006) Sick City, *The New Yorker*, 6th November 2006

[35] Human Development Report 2006 (2006) *Beyond scarcity: Power, poverty and the global water crisis*, United Nations Development Programme, UNDP, New York, USA

Woods R I, Watterson P A & Woodward J A (1988) The Causes of Rapid Infant Mortality Decline in England and Wales, 1861–1921, Part I, *Population Studies* 42 (3): 343–66.

Woods R I, Watterson P A & Woodward J A (1989) The Causes of Rapid Infant Mortality Decline in England and Wales, 1861–1921, Part II, *Population Studies* 43 (1): 113–32.

Szreter S (1997) Economic Growth, Disruption, Deprivation, Disease, and Death: On the Importance of the Politics of Public Health for Development, *Population and Development Review* 23 (4): 693–728.

Hassan J A (1985) The Growth and Impact of the British Water Industry in the Nineteenth Century, *The Economic History Review New Series*, 38 (4): 531–47

Bell F & Millward R (1998) Public Health Expenditures and Mortality in England and Wales, 1870–1914, *Continuity and Change* 13 (2): 221–49.

Halliday S (1999) *The Great Stink of London. Sir Joseph Bazalgette and the Cleansing of the Victorian Metropolis*, Phoenix Mill: Sutton Publishing.

Bryer H (2006) England and France in the Nineteenth Century, Issue Note commissioned for HDR 2006.

[36] Engineering Record (1903) Editorial: Sewage Pollution of Water Supplies, *Engineering Record*, 48, 117

[37] Tarr J A (1979) The separate vs. combined sewer problem: a case study in urban technology design choice, *Journal of Urban History*, 5(3), 308–339.

[38] *The Times*, 22nd June 1858, page 5

[39] Corfield W H (1875) *Sewerage and Sewage Utilization*, D Van Nostrand, New York, USA

[40] Sidwick J M (1976) A Brief History of Sewage Treatment, *Effluent and Water Treatment Journal*, April 1976, p195

[41] BMJ (1908) Royal Commission on Sewage Disposal, *British Medical Journal*, 7 November 1908, p 1447

[42] Royal Commission on Sewage Disposal (1908) *Methods of Treating and Disposing of Sewage*, Fifth Report, Royal Commission on Sewage Disposal, London, UK

[43] Beder S (1997) Technological Paradigms: The Case of Sewerage Engineering, *Technology Studies*, 4 (2), pp. 167–188

[44] Alleman J E (2005) The Genesis and Evolution of Activate Sludge Technology, http://www.elmhurst.org/DocumentView.aspx?DID=301 accessed 20th May 2011

[45] Niemczynowicz J (1997) State of the art in urban stormwater design and research, Paper at the Workshop and Inaugural Meeting of UNESCO Center for Humid Tropics Hydrology, Kuala Lumpur, Malaysia, November 12–14, 1997.

[46] http://www.trust.org/alertnet/news/unseasonal-rains-aggravate-cameroon-cholera-spread

[47] Owen D A Ll (2010) *Pinsent Masons Water Yearbook 2010-11*, 12th edition, Pinsent Masons, London UK
Owen D A Ll (2009) *Pinsent Masons Water Yearbook 2009-10*, 11th edition, Pinsent Masons, London UK
http://www.wssinfo.org/documents-links/documents/?tx_displaycontroller[type]=country_files
GWM 2011 (2010) Global Water Markets 2011, *Media Analytics*, Oxford, UK

[48] Solomon S (2010) *Water: The Epic Struggle for Wealth, Power, and Civilization*, Harper Collins, NY, USA

[49] WSP-SA / MoUD (2008) Technology Options for Urban Sanitation in India, Water and Sanitation Program-South Asia, The World Bank, Delhi, India

[50] WaterAid (2006) *Total sanitation in South Asia: the challenges ahead*, WaterAid, London, UK

[51] Hall D & Lobina E (2008) *Sewerage works: Public investment in sewers saves lives*, PSIRU, London, UK

[52] GNI data is from the World Development Indicators, World Bank, accessed 15th December, 2010 while the urban services data has been adapted from the author's water infrastructure database

[53] UNU-IWEH (2010) Sanitation as a Key to Global Health: Voices from the Field, UN University Institute for Water, Environment and

Health, Hamilton, Canada

[54] UNU-IWEH (2010) Sanitation as a Key to Global Health: Voices from the Field, UN University Institute for Water, Environment and Health, Hamilton, Canada

[55] Valfrey-Visser B & Schaub-Jones D (2008a) *Supporting private entrepreneurs to deliver public goods: Engagement strategies for sanitation entrepreneurs*, BPD, London, UK

[56] Valfrey-Visser B & Schaub-Jones D (2008b) *Engaging Sanitation Entrepreneurs: Supporting private entrepreneurs to deliver public goods*, BPD, London, UK

[57] EU (2009) Europa press release IP/09/369 'Cohesion Policy backs "green economy" for growth and long-term jobs in Europe' EU, Brussels, Belgium, 9 March 2009
EU (2005) Annual Report of the Cohesion Fund EU COM (2005) 544, EU, Brussels, Belgium
EU (1999–2004) Annual reports of the EU Cohesion Fund and Structural Fund, EU, Brussels, Belgium

[58] http://www.bbc.co.uk/blogs/panorama/2009/08/ accessed 3rd June 2011

[59] *New Scientist* (1987) Cleanup of beaches may be a waste of money, *New Scientist*, 21st may 1987, page 28
New Scientist (1985) The flotation that might sink, *New Scientist*, 18th April 1985, pages 16–19
New Scientist (1982) The sea as a sewer, *New Scientist*, 29th July 1982, page 294
New Scientist (1981) The unspeakable beaches of Britain, *New Scientist*, 16th July 1981, pages 139–43

[60] http://www.goodbeachguide.co.uk/search-results?beach_name=Eastbourne&wq_grade=0&postcode=&region=South+East&distance=10&lifeguards=0&facility_level=0&dog_restrictions=0&page_size=5&op=Search&form_build_id=form-88a76f61bdd7f5c5dc986ba521d7bc39&form_id=stbeach_search_long_form accessed 3rd June 2011

[61] Johnstone D W M (2011) Personal communication

[62] Clarke B (2006) Salmon are turning their back on Thames, *The Times*, 3rd April 2006
Pope F (2009) Why salmon chose to return to the Seine but not the Thames, *The Times*, 12th August 2009

[63] http://hfconservatives.typepad.com/residents_first/2011/04/fulham-says-no-to-super-sewer.html

[64] http://www.ft.com/cms/s/0/bf0f467a-5f4f-11de-93d1-00144feabdc0.html#axzz1LNJXOzIR

Chapter 5

Closing the Circle – Resource and Demand Management

The water cycle is about fresh water falling onto the land and treating and transporting used water to where it can eventually be usefully reclaimed. Despite the sheer scale of this ecosystem service, intensive use and indeed abuse mean that water availability is not always able to meet demand. The way the water cycle works can be mimicked beneficially by boosting the water flowing through the water networks via reclamation and additional treatment, mobilising seawater and wastewater, and optimising water usage. Treating water or wastewater is expensive, so such water needs to be properly valued. This calls for demand management, which is about going beyond providing water for the sake of providing water and controlling its usage so that demand does not exceed what is environmentally and economically sustainable.

Much of this is about innovation, a tricky subject for a business bludgeoned into conservatism, which simply has to take a fresh look at old challenges. Fortunately, we live in an era of almost unparalleled of innovation both in technologies and how they are applied.

Singapore's water independence

I have been to Singapore a few times over the past fifteen years, speaking at events such as the Singapore International Water Week and getting to know how the city-state manages its water resources. Its water management as well as the general level of economic development give cause for reflection, in particular, how the Public Utilities Board is allowed to manage the city-state's entire water cycle and its relationship with the government. Singapore has become a regional hub for advanced industry, especially in water management. Tour buses ferry people to and from the NEWater water reclamation facility; schoolchildren, pensioners and passers-by are guided through its potted history and futuristic treatment plant. The utility is working to a 50-year water management plan, it is state-owned but self-financing and operated by technocrats rather than politicians. Corruption is a folk memory and the private sector is welcome to build and operate new facilities on one condition – that they can push the boundaries of efficiency and deliver more for the utility at a lower cost. It is also unusual that politicians have consistently supported its aims, something which comes alive when you look at typewritten World Bank documents from the 1960s and 1970s where the planned improvements were in fact exceeded as the government grasped what they could achieve.[1]

Singapore is a tropical city-state blessed with year round rain (234 cm in an average year) but with just 710 Km^3 of land, it had 316 M^3 per person of internal renewable water resources available in 1965. Extreme water stress is defined as below 500 M^3 per person. When Singapore seceded from Malaysia that year, it had emerged from two long droughts, underlining its dependence on imported water from across the Straits of Johor. This left the country vulnerable to political interference with Malaysia's Prime Minister declaring that 'if Singapore's foreign policy is prejudicial to Malaysia's interests we could always bring pressure to bear on them by threatening to turn off the water in Johor.'[2] Forty-five years later, the first water import agreement is being allowed to lapse this year and the second agreement, which runs until 2061, may do so as well.

Singapore's population has grown from 1.9 million in 1965 to 5.1 million by 2010, during which time its water consumption rose fivefold. Sustainable water management is usually seen as a low priority in urban societies, due to a lack of public and political support for appropriate investment, let alone how their water usage affects water resources and about the applicability of various water sources. In Singapore's case, its need for self-sufficiency has resulted in a high degree of public understanding and acceptance.

Since its foundation in 1963, Singapore's Public Utilities Board has developed from a basic service to the most advanced water utility in the world and the first that actively manages its urban water resources in a way which replicates the natural water cycle. Currently, PUB sees water coming from 'Four National Taps': water from its own catchment, water imported from Johor, desalinated water and wastewater recovery (NEWater). From 2061 it is likely that water imports will be replaced by other taps, aided by water consumption avoided through network efficiency and demand management.

How does a developing economy achieve this? The first step was to ensure a reliable water service, one which people could trust. In the urban areas, 70–95% of households had piped water between 1916 and 1936 and in 1987 universal water provision was attained, providing continual supplies of potable water.[3] That is the public face of security of supplies, since when competence is the norm, public support can be gained, especially in a developing economy. Leakage and demand management followed, as unlined iron mains and connections were replaced from 1983–93 with all customers connected to water meters since 1983, and upgraded when needed to improve their accuracy. In 1984, distribution losses were 11%, a commendably low leakage rate by most standards, yet pared back to 5.3% by 2000 and 4.4–4.6% in 2008–09.

Public health and well-being are often interlinked but the extent by which one programme can benefit another is often under-appreciated. Until the 1970s, many rivers and streams were severely polluted through open defecation and pig and poultry farms, making their waters unfit for use.[4] The farms were closed and comprehensive sewerage and sewage treatment brought in, with 45% of people having household sewerage in 1965, rising to 75% by 1980 and all households since 1987. The rivers became water resources again, while improving the quality of life in a densely populated city state by enhancing their amenity value and making the riverside area desirable. In the long run, 90% of the island's land area is becoming available as a catchment.

As we saw in the previous chapter, sewerage is traditionally seen as getting foul wastewater treated and storm water back to the rivers and seas. Yet rainwater falling on cities is a resource in itself, and this was recognised by developing two separate sewerage systems, one for wastewater treatment and recovery and the other for channelling rainwater that falls onto buildings, roads and other hard standings either towards water catchments or water treatment facilities, so that they become a new resource rather than a nuisance. The potential for further reuse became clear, since 80–85% of water supplied by PUB in 1994 was already being collected in the sewerage system, and taken for treatment, showing how the water cycle was closing in.[5] Stormwater harvesting needs careful planning since not all run-offs are suitable; some industrial areas are too contaminated, while lake and estuary beds had to be remediated before they could be used.[6]

Three generations of sewage treatment have been deployed, moving from the basic local facilities of colonial-era expediency to six major secondary and tertiary treatment works which are now being phased out in favour of two central plants which in name and purpose have moved on from wastewater treatment to water reclamation. Thus wastewater is being seen as a resource rather than a waste.[7] The first large scale NEWater reclamation facility opened in 2010 after eleven years of trials and small scale plants. The water produced meets drinking water standards, but because of the inevitable squeamishness – I have drunk it and it is as agreeable as most premium bottled waters – drinking water from the plant come via the 'magic mile' whereby there is general public acceptance of abstracting treated effluent after it has been discharged into a surface water source such as a river or lake. Currently, 41,000 M^3 per day is being supplied for indirect potable use and 154,000 M^3 per day direct for non-potable use by industrial customers.

When confidence in both water and wastewater service provision is established then full cost recovery tariffs can follow. In the 1980s, domestic tariffs were typically subsidised by commercial and industrial customers with per capita consumption rising from 155 l/c/day in 1989 to 176 in 1994 as washing machines and power showers gained ground. Tariffs were reformed from 1997, removing all subsidies and raising taxes for water conservation and sewerage. Along with this, water use targets were set, with water use falling to 165 in 2002, and to 155 in 2009. Future targets are for 147 l/c/day by 2020 and 140 by 2030. There was a cost, as the average monthly bill rose from S$5 in 1980 to S$31 in 2000, but it fell to S$29.4 in 2004 due to lower consumption.[8] Encouragement for conservation was spurred by legislation. All new lavatories must be 4.5 litre low flush models since 1992 and after 2009, all domestic water appliances had to carry water efficiency labels, pushed forward by ever higher efficiency standards, with, for example, a household saving 2,600 litres per annum by using a 'three tick' rather than a 'one tick' low flush loo.

With full cost recovery, and network and efficiency measures in place, more costly and dramatic interventions such as desalination and wastewater recovery can be considered. The first desalination plant started in 2005; a second and larger desalination facility is expected to enter into service in 2013. Nothing is static and over the 50 years to 2060, total demand is forecast to double again driven by increased industrial demand, yet imported water is set to fall from 40% of resources to zero, with local catchments continuing to supply 20% of water, water reclamation rising from 30% to 50% and desalination from 10% to 30%.

There is another side to seeking self-sufficiency, which is a system's robustness. It is probable that Singapore's water resources will be amongst the most resilient globally in terms of dealing with challenges such as drought and climate change under the currently accepted scenarios.

Singapore did not seek to attain what it has achieved over the past 45 years in a single grand plan. Economic development has transformed the city-state's existence, its population and their expectations. Developments in desalination, water and wastewater treatment allowed for undreamt-of opportunities supported by decades of research and trials. What has been done is no accident: there is consistency in the execution of its infrastructure extension and enhancement, which stems from a single utility being in overall charge of the whole water cycle and having to answer to a single branch of the Government rather than various potentially conflicting interests.

When it comes to water management, you may consider Singapore as a major utility rather than a small country. The way it looks to its land as a catchment basin certainly fits in with this view.

From supply to demand

A supply-led approach can at worst simply be a case of throwing assets at a problem, while leaving the root causes untouched. This is especially important when you appreciate just how elastic various demand projections can turn out to be. Gareth Walker of the Oxford Water Futures group looked at 21 projections for future water supply needs in England and Wales made between 1949 and 2009 and plotted them against the actual amount put into supply over the years. In the mid 1970s, the utilities were supplying some 13 billion litres each day.[9] Two projections in that decade foresaw 28 billion litres a day being needed in ten to twenty years time, two more went for 24 billion and one plumped for twenty. In fact, supply peaked at 17 billion during those decades and today is back to 14 billion. What would have happened if those forecasts had been strictly followed? Billions of pounds would have been spent on new assets such as reservoirs, transfer pipes and treatment plants – none of which would have been needed.

Such projections, no matter how well intended, reflect a mechanistic, supply-led mindset, based on extrapolating from the past. A statist top-down model does not encourage innovation and indeed, there has never been a greater disconnect between the problems facing water management and the institutions which are meant to deal with it. Exceptions such as Singapore, where all aspects of water policy and water management were merged to eliminate the conflicts that typically dog the sector, are a pointer towards sustainable water management. There needs to be a fundamental shift towards appreciating the potential for innovation and demand management rather than throwing new assets at old problems.

Without the necessary political will, such a change is unlikely to happen. Management is usually constrained by factors beyond their control and so the sector's conservatism in response towards new challenges such as we are seeing can only change with suitable encouragement.

Metering has two sides

Back in 1989, a Midlands MP opposed to the privatisation of water and sewerage companies in England and Wales warned that once water was metered his constituents would face the threat of buttock bombardments as people started flinging faeces from high-rise flats rather than flush their loos. In fact, umbrella and hard hat sales stayed static and his prophecy slopped into history. This is a useful reminder about the level of debate we endure when it comes to water policy.

Metering and mindsets go hand in hand. There are two western economies without any domestic metering: Northern and Southern Ireland, where historical hang-ups mean that water is provided 'free' or paid through the rates and government subsidies. Such an attitude is rarely a good sign, with both these utilities having a poor record in service delivery and in Northern Ireland's case, weak corporate governance. 'If you cannot measure, you cannot manage' is something of a mantra in countries with a formally developed urban water service.[10] Indeed, in advanced countries, such as OECD members, outside the

British Isles and Norway, 60% metering is the minimum, with 100% being the norm, as also seen in many urban utilities in Central Europe, Africa, South East Asia and Latin America.[11]

Britain's utilities ought to join them. There was no metering in England & Wales in 1989, bar some field trials. Fairness on Tap, an alliance of conservation and water management groups, believes that metering in England should rise from 37% in 2010 to 80% by 2020 and from then on to universal metering. Metering is about informing customers about their water usage and enabling them to control it. When trials took place in England during the 1990s water demand fell by 10–15% after meters were installed, and up to 30% during the summer, when demand is at its highest.[12] That peak in demand is also when rivers are most vulnerable to over-abstraction. Looking at more recent data, average water consumption in unmetered households for Thames Water customers rose from 158 to 171 litres per person each day between 2008 and 2011, while falling from 144 to 141 for metered households during that time.[13] The relationship is not quite so simple, since just 28% of households are metered and when it comes to houses built before 1989 (after that date all households have meters installed), people can choose whichever option is the cheapest, which matters when someone likes to wave his garden hose.

Those who opt out of using a water meter are those who are most likely to lose from using them. They opt out because they can and therefore can continue to abuse water resources in an unsustainable manner. Once metered, they may well think again when their usage carries a price.

Pricing can be used to promote equity, efficiency and sustainability.[14] When the Massachusetts Water Resources Agency brought in fundamental tariff reform, it reduced water demand by 30% over two decades.[15] Logic would assume that you would either encourage water conservation by a fixed volumetric tariff (the more you use the more you pay) or an increasing block tariff (as consumption rises from necessity to luxury, you pay more per unit you use).[16] Uniform charges are used in the Netherlands, Sweden and France, while increasing or rising block tariffs are popular in Spain, Turkey and Japan, but in the USA, a third of bills use a decreasing block tariff, whereby the more water you use, the less you pay per unit of water – good news for the swimming pools of suburbia. Where meters remain the exception, a standard charge is in effect – an invitation to leave your taps running inside while leaks keep on leaking outside.

For the utility, metering is about knowing how much water is going in and out of a system, being able to identify when and where leakage is rising above an acceptable level and to do something about it, rather than chasing water seeping across pavements. Smart metering is next, which provides the customer and the utility with real-time data about how much water they are using and allows you to interrogate the data to see when abnormal usage is cropping up, from leaks to faulty equipment or simply to pinpoint where more water efficient goods can contribute the most.

Two early stage UK-based companies show how metering can be made to work for utilities and customers. Information and Performance Services specialises in real-time meter reading and analysis.[17] Its AutoChart package is a data handling and interrogation software, sending readings from a meter to a server, putting it onto the client's PC via the web, side-stepping sub-contract meter reading and allowing real time access to what is going on. This helps to pinpoint leaks, cutting them by 35% at 1,500 UK Ministry of Defence sites.[18] Senceive offers remote asset monitoring that 'replaces white vans with white boxes' by using an interlinked network of wireless connected sensors to provide continual information about the performance and condition of the network.[19] This is a lot cheaper than manually visiting sensors such as meters and overrides the danger of data transmission blockages which can happen, for example, every time a bus passes along a conventional network.

I have one concern about metering. If a utility gets more money when customers use more water and their bills are metered, what is their incentive to encourage customers to use less? Water utilities can benefit from lowered usage if driving down demand means that they do not need to develop new water supplies. Indeed, that can lower a utility's drain on nearby rivers, as well as curbing capital spending and customer bills. But if a utility is effectively incentivised to build additional assets and to benefit from these by what is known as the 'rate of return' on the utility's asset base, then there will be a constant regulatory pressure to build more assets rather than to curb demand as seen in England and Wales.

Leaking money and water

Leakage is a waste especially when you think of it as the loss of water that has often been specially transported and treated with our drinking from these taps in mind. Unlike metering, where common sense suggests universal coverage, leakage management is more about desired outcomes and local circumstances. Clearly, the less leakage there is, the better, but every utility reaches a point where money and resources spent on lowering leakage any further could be better used elsewhere.

There are two kinds of leakage a water utility needs to deal with: physical loss, where water is leaking from their supply pipes, and commercial loss, where it isn't paid for, leaking revenues. Unaccounted-for water is the proper term for leakage, the difference between what gets put in and what comes out. Non-revenue water is the difference between what the customer is billed and what they pay. There is no wriggle room here; illegal abstraction is theft, where the individual bleeds water and revenues from the community. The sheer potential from reduced leakage or the sheer wastage, depending on your point of view, was highlighted by the fact that halving water losses in low and middle income cities would provide enough water to satisfy a further 130 million people and improve cash-flow by $4 billion each year.[20] Other studies highlight the scale of losses,[21] with 42% leakage rates in major cities in Africa and Latin America. 39% losses are noted in Asia during the late 1990s, while a study by the Asian Development Bank[22] concluded that Asia loses 28.7 billion M^3 of water each year through leakage and the minimum value of this loss has to be $8.6 billion pa.

According to i2O, a leakage management company, 32 billion M^3 of water pa is lost through leakage worldwide, at a cost of £16 billion. Numbers collide, but whichever survey you take, curbing these losses would unleash both a cash and water flow that would go a long way towards dealing with lack of water access across the world.

In England and Wales, an act of political populism in 1998 meant that water companies were obliged to continue supplying water irrespective of their bills being paid or not. Under such circumstances, it is perfectly reasonable that debt counselling services advise people to stop paying their water bills straight away, and they do. Between 1999 and 2009, the amount of bad debt in England and Wales rose from £0.7 billion to £1.4 billion which, according to Ofwat, means that each of the poor fools who continue pay their bills are subsidising the non-payers to the tune of £12 every year.[23]

Pakistan might be a water-scarce country, but its seven largest water utilities had non-revenue rates of 32–46% when last surveyed for IB-Net the United Nations supported International Benchmarking Network for Water and Sanitation Utilities initiative. IB-Net abounds with insights and examples of good and bad practice. In Africa, you can choose between 59% unaccounted-for water in urban Mozambique and 44% in Tanzania or just 14% in Namibia, which is also a world leader in water reclamation.[24] Phnom Penh slashed its non-revenue water from 61% in 1996 to 6% in 2007, due to reforming billing, installing meters and a comprehensive revamp of its distribution network. Closer to home, during 2003, Scottish Water managed to lose 43% of its water compared with 25% in Wales, 26% in the distinctly drier

Thames Water area and 14% for Southern Water. Given that south eastern England has the same per capita water availability as Syria, with 40% of all rainfall already used and its burgeoning population, leakage is already something of a hot potato and is going to become an ever more critical issue there.

In a perfect world, there would be no water loss, but, in reality, utilities seek a balance between current real losses and unavoidable real losses. For example, the more connections and joints there are, the more opportunities arise for water to ease between them. Higher water pressure, especially when serving high-rise buildings makes this worse. This is the background leakage and the question is what level is acceptable and affordable. The American Water Works Association recommends that 10% unaccounted-for water is a benchmark for the well-managed utility, with action definitely called for when it goes above 25%. I would say 10–15% is a sensible target, certainly seeking to stay under 15% and going under 10% when circumstances and needs allow.

As we saw in Singapore, distribution losses indeed matter more where there is less water. At 10.8%, the leakage in Sydney's water in 2004 was unacceptable in the context of its supplies. By 2010, leakage was down to 6.6%, a saving of 66 million litres a day since 1999. In Riyadh, leakage has been ignored to the point where 1.1 million M^3 of water is lost every day, the equivalent of nine major desalination plants working flat out to meet this unnecessary demand.[25] Profligate water use in Saudi Arabia was the unsustainable legacy of a subsidy culture whereby tariffs covered one tenth of the utility's operating costs and the networks were not maintained, rapidly becoming decrepit. Another arm of the subsidy culture spent eye-watering sums on desalination plants, piping water 450 Km to Riyadh where 31% of the water promptly drained away through leaks. A sieve is a pretty silly way to convey water at the best of times, let alone when that water is costly and scarce. By spending $400 million in basic network repairs in Riyadh, $2.1 billion can be saved over twenty years in avoiding the need for new desalination plants, another example of improved demand management obviating the need for excessive infrastructure and spending.

All forms of wastage can be reduced and have been reduced, ranging from grinding down unaccounted-for water in Lusaka, Zambia from 52% in 2001 to 25% two years later to a one year project that saw leakage in Zagazig, Egypt fall from 28% to 10%.[26] Over the longer term, even more dramatic progress can be made. Since taking over the eastern portion of Manila's water and sewerage services in 1997, Manila Water has reduced unaccounted-for water from 63% to 11% by the end of 2010, with the water saved going to the newly served as coverage rose from 3.0 million to 6.1 million during this time, with the greatest outreach being in the poorest areas.

Billing and bill collection can be a matter of attitude. Removing illegal abstractions made good sense in Phnom Penh and Manila, especially when replaced by a reliable and better quality service. Uganda's National Water and Sewerage Corporation has been transformed from a litany of financial and physical loss to a bastion of good management through a systemic linking of cost recovery, bill collection and improved service delivery.[27] Between

1998 and 2010, connections rose nearly fivefold, non-revenue water falling from 60% to 33% and bill payments rising from 60% to 100%.[28] At a time of unprecedented urbanisation, service coverage rose from 48% to 74% without having to increase production capacity, while being both profitable and able to charge just 9% of what the water vendors do. People will readily pay for services which they can trust and are seen as offering good value.

A shift in techniques is taking the detection of leaks from those that reach the surface to those that do not. The latter are not so spectacular, but they cause the long-term losses. That means technology moving from listening out for distant gurgles (a wooden pole is used to this day) to sending remote monitors through the network to look out for cracks and using metering to highlight anomalous water use. Here, Norway's Breivoll uses acoustic resonance technology via remote controlled 'cars' that travel down the pipes, to generate a picture of the actual condition of a pipe.[29] Some, which are 100 years old, are not overdue for replacement, as it turns out they have at least a further century of useful life left in them. Techniques like this can allow a shift from replacing assets because you feel you ought to, to doing so when it matters. Where appropriate cast-iron piped can be replaced with ductile iron, steel or plastic pipes. Monitoring and managing pressure within a pipe network reduces leakage, as what goes in within a system is going to find the nearest point for going out. In Sydney's case, 21% of the leakage reduction came from more active pressure management rather than repairing pipes.[30]

As well as many companies offering remote asset monitoring, companies such as i2O[31] of the UK seek to lower leakage by ensuring there are no parts within the distribution network which have a higher than necessary water pressure. Trails with Severn Trent in the UK claim a 26% reduction in leakage with a four month payback time. Pressure management irons out differences during the daily demand cycle and ensures optimal pressure across the system. Meanwhile Austria's Sanipor offers a no dig, in situ sewerage rehabilitation service for covering the other side of leakage. Sewerage rehabilitation is usually a costly process; and Sanipor believes it can lower the cost of a typical sewer rehabilitation task from $34,000 to $15,000.[32] This is a pragmatic approach which has been approved for use in Germany, the USA and UK. A solution is pumped in, which fills up holes outside the pipe and after a second solution is pumped in and solidifies the plug into a concrete-like matrix.

How far can we go? Berlin Wasser, serving Germany's freshly reunified Berlin has a modern and compact water distribution system, with distribution losses of 2.9% in 2008. This suggests that a long-term ideal range for distribution losses in Singapore would be 2.5–4.0%. Distribution losses of 6% to 14% occur amongst utilities seen as good performers in England, the Netherlands and Germany.

Water efficiency – the demands of demand management

A water meter, full cost recovery and a properly itemised bill can do wonderful things. Taking a hard look at how and why your water bill adds up can be better still, especially as you get to grips with the individual drivers that make us use more water.

Simply by launching a $250 million lavatory rebate programme, swapping the traditional 5–6 gallon (22–25 litres) flush cisterns with a saner 1.5 gallon (7 litres) flush, between 1992 and 1997 New York was able to drive water consumption down by 20%, avoiding the need for new water supplies.[33] Why the flush had to be so large in the first place is another story. This might be basic compared with Singapore, but the scale of the saving here is telling. Anyway, Manhattan Island has no known need to seek self-sufficiency, since as we saw earlier its upstate catchment preservation programme is a pretty popular one with its neighbours.

A study in water efficient Europe shows the scope for cutting consumption simply through replacing your household goods with more efficient versions.[34]

Lowering water use through water efficiency

(Litres per household/day)	Standard	Efficient	Reduction
Lavatory	57–87	39	32–52%
Shower	45–54	43	3–44%
Bath	71	53	26%
Taps	10	8.5	15%
Washing machine	26	17.4–19.6	25–33%
Dishwasher	8.7	5.2–6.1	30–40%
Total	237–280	167–169	29–41%

Here it is the hardware that is doing the job and human habits do not need to change. Admittedly, the bath will be a slimmer proposition and the power shower will not be the deluge of old but, with hot water, the payback is redoubled, because of the lower electricity bills. As water is electricity's poor cousin when it comes to utilities, it is little surprise that when water efficiency means energy efficiency, people really start to notice. Heating water accounts for 30–50% of household energy use, meaning that minimising the use of hot water in showers, dish and clothes washers as well as through smart metering offers much more than simply saving water. Since two impacts are being reduced by one device, it ought to incentivise both water and power utilities and their customers.[35] The water-energy nexus works in many ways. Perhaps the most effective method of encouraging water conservation is to highlight the energy saved. People notice their electricity bills and these are invariably metered.

Strategies for implementing demand management continue to evolve as opportunities are identified. Pachube is a US company that is developing Internet software that warns people to avoid flushing water in New York when the city's combined sewer system is close to overflowing by monitoring levels at overflow locations during heavy rainfall and relaying this to customers who can then chose to delay running a bath or starting a wash cycle.[36] At the corporate level, the Aqueduct Alliance, which consists of seven leading US companies and the World Resources Institute, is rolling out a database of river-basin water availability which they hope will allow companies to take a more rational approach towards where they place water intensive facilities or indeed make them more water efficient.[37]

A reasonable domestic water consumption target[38] would be 120–140 l / cap / day as 120–128 l / cap / day is already the norm in parts of Germany and the Netherlands. As ever, nothing is quite so simple in water and the lowering of water consumption has had two unintended consequences. Longer residence times of drinking water within German distribution networks had meant that treatment standards have needed to be enhanced to prevent bacterial contamination building up and it has become even more of a challenge to flush sewage down the system when the flow is low. Sometimes a little profligacy has its charms when it comes to moving matters along.

Eco Cities and water footprints

Masdar City is a proposed $22 billion 'Eco City' for 45–50,000 people in Abu Dhabi, UAE. All of the city's primary water will come via desalination. A proposed 60% reduction in water usage against the norm in the region equates to residential usage to 180 l/c/day, which remains high by European standards, especially given the scarcity of water in the region. Through further demand management and water recycling and recovery, the developers aim to reduce net water usage to 70 l/cap/day. The 'Eco City' concept is based on no net depletion of local water resources, nor any appreciable nutrient loading downstream. Solar powered desalination and effluent powered water reuse are plugged in, integrated with demand management and greywater reuse. It is an attractive concept, but with an implied cost of $0.45–0.50 million per person for the whole package, there remains a gap between theory and practice. Since the 'Eco City' term was coined in 1987, quite a number of housing projects and proposed cities have adopted the term, but none has met all the criteria for water, energy and waste management.

To dismiss such aims as 'ego cities' would be cruel, given that there is a good idea behind them. Perhaps 'Eco Cities' are wish lists for a fully funded, fully realised future when water scarce communities can manage their water footprints to the point where they do not affect their surrounds. Cities and urban societies have two water footprints. One lies upstream and uphill, for abstracting water either to quench their thirst or for growing their food and another, lying downhill and downstream for nutrient and pollution loading.

I am a bit cautious about water footprinting as it currently stands, because the numbers cited can be misleading at best and there are various motives for promoting them. For example, intensively raised cattle fed on a grain-based diet have a far higher water footprint than cattle raised on natural grassland which has a marginal value for crops. Almost invariably, it is the grain-fed beef that is used on footprint calculations since this highlights a burger's higher water intensity than a nut cutlet.

In time, as numbers, definitions and context are clarified, water footprinting can become a powerful tool for looking at what drives individual and community water use. I suspect it will start with how much water is taken out and then after moving to what is discharged will go into the energy – and indeed carbon – intensity of water and wastewater. The effect of this will be to focus people's minds on minimising the need to abstract fresh water and to maximise the energy that can be recovered from wastewater. Economic as well as environmental sustainability means that water utilities need to be assessed for their carbon impact and to be incentivised to minimise it, with the life cycle impact of new assets forming part of the procurement process.

Is desalination sustainable?

Many aspects of our history revolve round extracting salt from the shores while today humanity needs to extract water from the same seas.[39] Not only is humanity increasing and urbanising, it is shifting towards coastal areas. According to Colombia University's Centre for Climate Systems Research the number of people living within 60 miles of the coast will rise by 35% between 1995 and 2025 to 2.75 billion, while 20 of the world's 30 largest cities are defined as coastal, areas where water scarcity or stress is more pronounced.[40]

The marriage of salt and water has mingled in our imaginations for a long time. The first written speculations we have on the nature of the hydrologic cycle were written by the ancient Greeks.[41] The most influential of the Hellenes was Aristotle (384–322 B.C.). He understood the process of distillation and knew that rainwater could be generated by evaporation of seawater:

Salt water when it turns into vapor becomes sweet, and the vapor does not form salt water when it evaporates again. This I know by experiment

Aristotle, *Meteorologica*

Although Aristotle understood the process by which seawater is transferred to the land, he struggled with the question of how streams can continue to flow for many weeks in the absence of rain. In one edition of his *Meteorologica*, he does allude to what we could regard as a porous medium even if reverse osmosis was one step too far.

Using membranes to mobilise sea and brackish water resources to complement surface and groundwater supplies has transformed desalination from a fringe activity for petro-plutocrat funded distillation plants into a globally significant water resource. Because desalination is a particularly costly and energy intensive way of obtaining water, it can be a controversial one but it is one that cannot be ignored. Despite being a relatively new technology, reverse osmosis accounts for 61% of the current global installed desalination capacity of 59.9 million M^3 per day. This reflects their appreciably lower costs while now delivering comparable water quality and reliability. The typical cost for Multi Stage Flash distillation systems is \$2.00–6.00 / M^3 of water desalinated over their operating life, against \$0.45–0.80 / M^3 for a reverse osmosis membrane facility.

Ethics and desalination

There are three areas of concern regarding the ethics of desalination: where it fits in the current debates about the politics and economics of water provision; the need to invest in mobilising new resources or in ensuring current resources are used more efficiently and how environmentally sustainable desalination is. The World Bank regards desalination as a last resort and not to be used where a utility is poorly managed while the WWF wants desalination to have the equivalent of the World Commission on Dams in effect calling for a moratorium on major projects.[42]

Water as an 'uncooperative commodity' has been adopted as a proxy for wider political and social dissent in recent decades. Desalination has inevitably been drawn into this, often because of its relatively high profile. Thames Water (then owned by Germany's RWE AG) sought to build a 140,000 M^3/day brackish water RO facility for 900,000 people at Beckton in East London in 2004. This was subjected to a court challenge in 2005 by the Mayor of London on the basis of the project's carbon emissions and arguing that it ignored leakage management (a not immaterial 915,000 M^3 of distribution losses per day in 2004–05). After the election of a new Mayor in 2008, the legal challenge was withdrawn in return for the facility being powered by recycled biofuels, or more prosaically, used oil from London's chippies. There is a delicious irony here, since the job could have been done better and cheaper by using wastewater recovery, but it is hard to imagine any local politician in the UK allowing that.

Progress has been made on mitigating the environmental impact of desalination. Inlet flow management can minimise the potential for taking up aquatic life, while brine discharge can either be avoided (discharge to surface sites, but this in turn had land / groundwater contamination concerns or more usefully, the prospect of brine recovery) and the use of diffuse discharge or brine blending / mixing zones can ease the discharge impact, provided suitable locating and monitoring is carried out from the outset.

Energy intensity remains the greatest challenge. Because of the size and technology-driven nature of the market ($10.8 billion went on building desalination plant in 2010 according to Global Water Intelligence), it is a competitive and innovative one, especially when it comes to dragging energy consumption down. Improving the efficiency of membranes, allied with new technologies (membrane-condensation, waste heat, forward osmosis, aquaporins and nanotubes) and energy recovery systems, indicates that there is more scope to minimise their impact, but reverse osmosis is an intrinsically energy intensive approach.[43]

Under such conditions, marginal advances can create major market opportunities. In 2011, Aquatech of the USA gained a desalination contract in the UAE using a high-retention nano-membrane rather than a conventional reverse osmosis membrane.[44] While the best reverse osmosis plants consume 5.2 kW hours per cubic metre of water produced, the Aquatech facility's energy requirements will be 3.14 kWh / m^3 more than making up for the plant's higher construction cost. Distillation may be making a comeback through using solar energy to directly vaporise seawater and collecting the condensate. Korea's Doosan won a contract to develop such a mid-sized multiple effect distillation facility for Saudi Arabia's Saline Water Conversion Corporation in 2011, using renewable energy sources offers the potential for carbon neutrality.[45] To date, examples include cooking waste biofuels (Beckton, UK) and photovoltaic energy (Perth, Australia) but solar power, like wind, depends on places where conditions are right. Tide and wave energy may offer potential, especially for smaller plants, but these are at the development stage.

Finally, where wastewater is being treated in the vicinity, sludge to energy can be a significant energy source.

In addition, desalination can in fact counter the effects of climate change. By supplementing stressed water resources it can be switched on to prevent river over-abstraction and respond to greater seasonal change (drier summers and wetter winters) as well as addressing episodic droughts. It also provides planners with a more nuanced responsiveness to population change. The modular nature of many techniques means that capacity can be altered to circumstances and for the facility to be used only when required.

Taking water management seriously

The high relative cost of desalination means that water economics are more taken into account than elsewhere. In a commercial desalination project, desalinated water is sold by the operator to the municipality on a commercial basis and the municipality is in turn more likely to seek to minimise distribution losses and encourage water efficiency in order to optimise the use of a costly resource. This usually means full cost recovery pricing at the customer rather than the utility level, along with water metering and progressive (rising block) tariff structures with affordability support where appropriate along with the whole range of demand management tools, as seen in Singapore. Treating desalinated water as a commodity encourages its sustainable management through ensuring its optimal beneficial use at the lowest achievable cost. This in turn may become a tool for regulation-led demand management, whereby utilities are incentivised to encourage lower water demand.

Turning wastewater into a resource

Traditionally, we have dismissed wastewater as a waste. Yet a better way of looking at wastewater is to consider its potential as a vast conduit of embedded resources awaiting beneficial reuse and going towards making the sector carbon neutral through wastewater mining. When it comes to nutrients, metals and energy, this is broadly accepted, but water recovery remains more challenging.

Attempts to recycle water have to deal with politicians and populism. After four years of rising water restrictions, the Australian City of Toowoomba held a referendum in 2006 on a water recycling project after special interest groups opposed the proposal.[46] CADS or 'citizens against drinking sewage' has a track record of fighting water reclamation plans across Australia and weighed in with the fear factor. Blog Toowoomba raised the tone of the debate with headlines such as 'I don't want to die mummy', while others worried that the town would have to be renamed Poowoomba.[47] Perhaps inevitably, 62% voted against the A$68 million project, and full water supplies were only fully restored in 2010 through an A$185 million water diversion scheme. Yet as seen in Singapore and also in Windhoek, Namibia, where they have been using direct potable water recovery for a decade and no illnesses arising from this water have been identified during that time, water recovery can be an integral part of the wastewater treatment process and it can be reused, either for potable or non-potable purposes.[48]

Sludge to energy is an important area because of the greenhouse gas impact of methane. Optimising the conversion of sludge to methane which is then used to generate electricity means that a significant proportion of a wastewater treatment plant's energy needs can be internally sourced, doubly lowering the processes' carbon footprint. Power generated from sludge reached 2,939 GWh in 2010 worldwide, with 35% of this coming from Germany.[49] In theory, wastewater contains ten times the energy it takes to treat it. A realistic target could be generating four times the amount used during treatment. This is very much a work in progress. In Germany, trails found that co-fermenting municipal wastewater generated enough energy at the on-site gas engines to cover 113% of the energy the treatment facility used.[50] Taking things further, trials are looking at using wastewater as a feedstock for algae which can become a biofuel along with urine being a fuel for microbial fuel cells,[51] which would provide an energy source for remote houses and villages.[52]

Recent research at Newcastle University has found that wastewater has an internal chemical energy of 7.6 kilojoules per litre, 20% more than had previously been assumed and this does not take into account the energy stored in industrial and commercial wastewater.[53] For seven billion people, this works out as 2.7–5.4 exajoules (ten to the power of eighteen joules) of energy or 0.6–1.1% of global energy consumption. Not much in itself, but given that the water sector contributes directly 1% of the UK's greenhouse gases[54] and indirectly 5% in the USA[55] (and more in the UK) when you take into account water heating, then making use of this embedded energy makes good sense. Getting that energy beyond the wastewater treatment works will involve new approaches such as microbial fuel and electrolysis cells. But the sludge the team investigated is being put to good use, with Northumbrian Water's Bran Sands wastewater facility cutting their energy needs by 92% after investing £33 million in advanced anaerobic digestion and related sludge to energy and heat recovery systems, saving the company £5 million a year.[56]

People do take notice of their electricity bills. Between 2003–04 and 2009–10, the jump in power costs transformed the way people regarded energy as a cost for water and wastewater services. In 2003–04, power costs accounted for 8% of the water and wastewater sector's operating costs in England and Wales, but by 2009–10 this was 13%, although falling from 10% of sludge treatment and disposal costs to 6%. The overall rise reflected the increase in power costs[57] and the growing demand for pumping water and wastewater. The lower costs for sludge treatment and disposal are due to the progress being made in sludge to energy to date as shown in the example by Thames Water below. This is why it makes good sense to look at sewage as an embedded energy resource. Thames Water has 349 wastewater treatment works, 22 of which generated renewable energy in 2008–09, accounting for 14% of the company's total energy consumption. Two works, Beckton and Crossness, which serve 5.5 million people and generate energy from burning partially dried sludge while twenty smaller facilities used anaerobic digestion of sludge to generate methane for combined heat and power burning. The energy generated saved Thames £15 million in 2008–09 and they are aiming for a 20% drop in greenhouse gas generation from 1990 levels by

2020 despite having a significantly higher energy demand.

Other sludge applications need to put the sludge through further processing, such as sludge-to-energy or composted (on its own or with green waste such as grass cuttings) and pelletised for various horticultural applications. Attempts to make sludge into food in Japan in the early 1990s worked in terms of their texture and appearance but got nowhere with regards to taste and smell and were quietly forgotten as were other projects such as turning sludge into jewellery and paving stones. Encos, a Leeds university spin-off company is currently developing building bricks from various waste by-products including ash from sewage sludge incineration.[58]

The nutrient cycles operate more slowly than the water cycle and sources of fertiliser are limited and non replenishable. It is therefore essential that as much of these nutrients are recovered as possible, especially since as we have already seen, their discharge into lakes and rivers is a source of harm. James Barnard has been working at biological nutrient recovery systems for three decades. During that time, he has seen the preferred treatment shift from biological to chemical and back again. He notes that half the world's population cannot afford fertilisers and that urine goes some way to making up for this, and that urine recovery ought to be the norm, whether through separate collection systems or recovery at the treatment plant.

10% of irrigated water worldwide comes from wastewater but it is usually untreated or partially treated wastewater[59] which is also a source of disease. The scale of agriculture which uses such wastewater is immense. Some 200 million farmers and their families use such water worldwide, providing vegetables for 700 million people. There does not appear to be any reliable data on what this does for the consumers' health, let alone that of the farmers' but it must be appreciable. Rather than straight discharge, wastewater used ought to be subjected to some form of disinfection before discharge.

Wastewater recovery from treated water remains at an earlier stage of development. Suez Environnement estimates that in 2008 some 368 cubic kilometres of wastewater is collected in sewerage systems each year, of which 160 are treated and 7.1 is reclaimed, 4.5% of the total treated.[60] But the costs are improving, especially as energy efficiency improves. GE believes that while energy and other costs have been pushing up the cost of 'traditional' water treatment from $0.3 per M3 in 1996 to $0.6 by 2010, they have fallen from $0.7 to $0.4 for reuse during this time.[61] That may be an over-simplification, as reuse needs energy as well, but the point about such approaches becoming increasingly viable is a valid one.

Water reuse costs and applications ($ per M3)

Process	Application	Cost
None	Informal discharge	0.00
Secondary	Restricted irrigation	0.01–0.15
Tertiary	Landscape	0.16–0.36
Tertiary	Industrial process water	0.18–0.60
Advanced	Groundwater recharge	0.45–1.20
Internal	Zero discharge	0.80–1.50

Just as the more advanced the application is, the greater the cost involved, the same goes for the energy needs. Perhaps one lesson here is to focus on the most appropriate applications for the particular area. In Namibia, Singapore, the UAE and even Saudi Arabia, cost recovery has been achieved by fixing the price with the user and to fix the quality of the water delivered to this.[62]

Driving down water energy needs (kW hours per M3)

	Current	Potential
Desalination – reverse osmosis	3.0–5.0	1.5–2.5
Reclamation – potable	0.4–1.4	0.3–0.4
Reclamation – non-potable	0.8–1.0	

Since the economics of wastewater recovery favours more advanced treatment, there is some scope for bypass secondary treatment and going straight to tertiary, with resource recovery in mind. This could be a developing country revolution, making more advanced treatment more feasible.

Stormwater, greywater and rainwater harvesting

Indirect water reuse using the 'magic mile' takes place without it being noticed, so that in theory, water can be used, treated and recovered seven to eleven times down rivers such as the Thames and the Rhine with most people never appreciating that this happens every day. Likewise, using rainwater harvesting has the potential to bring average water consumption down from 150 l/c/day to 105 l/c/day and when combined with grey water recovery for flushing lavatories, to 70–80 l/c/day.

In the UK, grey water usage remains at the development stage. The only major housing development where it is used is the BedZED (Beddington Zero Energy Development, Sutton) project, while smaller projects have been seen in recent years, along with at offices, schools and community centres. In the USA, the Uniform Plumbing Code, which is used by some States, actually prohibits the use of greywater inside a house, while the International Plumbing Code which is used in parts of the USA, Europe and Australia does. This can occur at the building level or as we have seen in Singapore, at the street level. According to the UK Rainwater Harvesting Association 400 properties pa were being connected in the UK in 2004–05, but this generally remains an ad hoc and somewhat informal market.

Cleaning up with SUDS

Cities do not have to be hard engineered seas of tarmac and concrete. Sustainable Urban Drainage Schemes (SUDS) seek to replicate the absorptive capacity of natural systems and their ability to deal with episodic rainfall. When you have more trees in cities, they absorb rainwater as well as providing shade. Inner city areas can have 'green' walls and roofs, along with porous pavement areas and pocket parks. The suburbs offer space for small ponds and roadside swales, while the urban outskirts ought to offer large ponds, reed beds and constructed wetlands and wherever possible, river restoration. Nature needs space and space is at a premium in large cities and so building such schemes back into an over-built city will not be cheap. The real power of SUDS lies in preserving and enhancing suitable habitats and designing them into new developments. Likewise, reed beds treat wastewater and where sewage is discharged onto an estuary: Biomatrix Water is a typical example of a company seeking to replicate this effect through land and water based treatment systems using units holding plants and settling zones to reduce the pollution loading of industrial and municipal effluent and to minimise sludge generation.[63]

Integrated Water Resources Management (IWRM) and its bigger-picture child Integrated River-Basin Water Resources Management (IRBWRM or IRBM for short) emerged from the 1977 Mar del Plata Conference to 1992's Rio and Dublin conferences on water and sustainable development.[64] While the word 'integrated' reflects the need to look at the way supply and demand along with effluent prevention and treatment ought to be balanced, it is clear that regulation and management are usually anything but integrated, with vested interests from political taboos about rainwater harvesting or bowing to lobbying about pollution sources managing to overrule any practical implementation. This is particularly the case when a catchment degrades and additional treatment is needed. Piecemeal policy is cheap to implement but expensive to treat.

SUDS and IWRM have a great deal to commend them, but they are radical options when compared with conventional approaches towards managing water and wastewater and the sector remains timidly conservative.

As DEFRA highlighted in its 2007 consideration of the Thames drainage problem, their potential remains perhaps under-appreciated.[65]

Conclusions – a world of innovation

Control those taplings and declare for sense

John Dryden, *Cleomenes*

In a talk at Oxford's 'Third Water and Membranes Research Event' in 2010 Richard Bowen looked at 'engineering and the idea of justice' and the potential for innovation to add to human happiness.[66] His approach in part stems from Amartya Sen's 2009 book *The Idea of Justice* which proposes that justice can transcend theory and be an act of 'practical reasoning' so that the obligations that power carries have three salient outcomes; if it is good ('justice enhancing') and it is freely doable ('feasible'), then that person in a position of power needs to take it seriously as justice depends on capabilities as well as will.[67] Professor Bowen depicts practical as 'everything works, don't know why' and theoretical as 'nothing works, do know why' which elegantly sums up our understanding of water and sewerage networks. The engineer ought to look beyond the technical achievement towards the humane, or as Bowen puts it, a scientist or technician may be happy to develop or to make a device while an engineer will want to see a device operate and preferably to do so commercially. He cites Sydney Loeb, who, during a decade as an academic engineer in California discovered the possibilities for reverse osmosis in desalination in 1959, developed the first commercial plant for brackish groundwater at Coalinga in 1965 and the first seawater facility at La Jolla in 1969. Sen has been a force for the water good before, as his earlier writings formed the philosophical pillar for the United Nations' Millennium Development Goals of 2000.

Innovation comes with a price – the price of funding it, developing it and making the message heard. CleanTech is the catch-all phrase for environmental / sustainable technology, usually covering areas such as renewable energy, smart metering, monitoring and control systems and waste reclamation as well as water and wastewater. Money rarely hurts, especially when you are developing new products and processes. This is where the Venture Capital industry comes in, providing various funding 'rounds' from conceptualisation to commercialisation. It's a big business, with $30–40 billion being invested in recent years from health care to software and inevitably, the Internet. CleanTech has become a major component of VC in recent years, with $6–9 billion invested annually, but at between $124 million and $257 million a year, water accounts for just 2–3% of the global CleanTech venture capital market.[68] Optimism abounds, with forecasts of water's share growing to 15% in the near future[69] and 24% of CleanTech investors[70] seeing it as a priority, but longer-term followers of water finance may remain a bit more cautious.

There is room for optimism, as long as the funding can be arranged. The pace of innovation in recent years has been greater than ever during the history of water and wastewater management. It is interesting to note that in the USA, innovation takes place at the local and state level rather than the federal level. In the Netherlands, Germany and Singapore, often innovation is driven by Governments encouraging it through new standards and strictures. In general, I suspect that the future for innovation and best practice is at the city, city-state and water basin level: a case of local responses to wider drivers.

Sometimes innovation consists of mobilising the apparently mundane. Banana peel removes over 65% of heavy metals from contaminated water in less than 40 minutes and other plant wastes look like having similar properties.[71] This way, instead of having to import high-cost hardware, developing countries may be able to use materials that are local and plentiful. For a conservative business, innovation is everywhere. Water recovery is being made less energy intensive through using a membrane bioreactor (the great 'disruptive' water and wastewater treatment breakthrough of two decades ago) instead of various sedimentation and filtration processes. In Singapore, seven years of field trials identified eight ways of optimising the process which drove down the energy needed to recover a cubic metre of water from 1.3 kilowatt hours to 0.37, starting with three pilot plants and ending up with a fully operational system.[72] The greatest opportunities lie when you plan infrastructure for incipient urbanisation. Anticipate and locate is a useful mantra here.

Of the 120–700 litres each piped-in urban dweller consumes every day, perhaps 10–20 of these actually needs to be fully potable for drinking water and for preparing food. All that water sluicing through each system that gets used for washing, flushing or simply watering things has been appreciably over-treated for its intended use. In the USA, a lot of water is in fact consumed either as bottled water or as beverages ranging from homeopathic coffee to carbonated sugar substrates. One of the reasons for this is that the drinking water tends to be heavily chlorinated and tastes that way, which also explains the thriving market in 'point of use' domestic water filters and bottled water. So the 1–3% of water that is consumed is generally safe but can taste grim, while the other 97–99% is arguably over treated. How far can you go when decentralising services? If non-potable water was delivered to the house and you had a point of use filter at the drinking water tap, you could avoid unnecessary treatment for the 90–95% of household water that does not need the potable polish. But what would the energy, operating and financial costs be of maintaining those filters and indeed, would they be maintained?

This calls for an open-minded approach to the potential for innovation and regarding corporate innovation as a potential source of succour rather than a threat as innovation is required right now right across the water cycle. What about higher tariffs during drier seasons or metering domestic and commercial wastewater? Then there are more radical approaches towards treating water such as 349Q, a company I was shown in 2009, which proposed using RNA interference in a bio-nanofilter to prevent gene expression by bacteria and viruses. Bluewater Bio's HYBACs sludge digesters shrink the land needed to treat sewage, which makes it possible to treat wastewater in densely populated areas. By cutting the utility's operating and capital costs, HYBACs also meets the golden rule for sustainable innovation; better performance at a better price.[73] Meanwhile, some processes such as clothes washing are seeking to go 'post water' but the practical obstacles here remain considerable. Industrial water use could be driven right down by internalising water use rather than abstracting and discharging. This closed cycle approach is set to become increasingly important as utilities make industrial clients pay a full fee for bulk water usage, a fee that may make some approaches to water usage economically as well as environmentally unsustainable. When we properly appreciate the cost of abusing the water cycle, we will know the proper price for water management.

---

[1] International Bank for Reconstruction and Development (1968) Report – on a proposed loan to the Republic of Singapore on a proposed sewerage project, Report P-612, World Bank, Washington DC, USA
International Bank for Reconstruction and Development (1973) Report – on a proposed loan to the Republic of Singapore on a proposed sewerage project, Report P-1280, World Bank, Washington DC, USA

[2] Luan I O B (2010) Singapore Water Management Policies and Practices, *Water Resources Development*, 26, 1, 65–80, March 2010

[3] Otaki Y (2004) Water systems and urban sanitation in Tokyo and Singapore during the 19th to 20th centuries, Paper read at the 6th International Summer Academy on Technology Studies, Deutschlandsberg, Austria, July 11–17, 2004

[4] Choong K Y (2001) Natural Resource Management and Environmental Security in Southeast Asia: Case Study of Clean Water Supplies in Singapore, Paper No 15, Institute of Defence and Strategic Studies, Singapore

[5] Choong K Y (2001) Natural Resource Management and Environmental Security in Southeast Asia: Case Study of Clean Water Supplies in Singapore, Paper No 15, Institute of Defence and Strategic Studies, Singapore

[6] Lim M H at al (2011) Urban Stormwater Harvesting: a Valuable Source of Water Supply in Singapore. Paper presented at the Singapore International Water Week, Singapore, 7th July 2011

[7] Gee T E, Stratus (1992) Trends of Waste Management in Singapore in Chua T-E & Garces L R eds. (1992) Waste management in the

coastal areas of the ASEAN region: roles of governments, banking institutions, donor agencies, private sector and communities, Proceedings of the conference on Waste Management in the Coastal Areas of the ASEAN Region, Singapore, 28–30 June 1991

Khoo T C (2009) Singapore Water: Yesterday, Today and Tomorrow, in Biswas, A K et al (eds.) *Water Management in 2020 and Beyond,* Springer-Verlag, Berlin, Germany

Ong I B L (2010) Singapore Water Management Policies and Practices, *International Journal of Water Resources Development*, 26, 1, 65–80

[8] Tortajada C (2006) Water Management in Singapore, *Water Resources Development*, Vol. 22, No. 2, 227–240, June 2006

[9] Walker G (2011) Models of domestic demand in the UK water sector – science or discourse? Workshop on Water Pricing and Roles of Public and Private Sectors in Efficient Urban Water Management, Granada, Spain, 9–11th May 2011.

[10] Kayaga S (2008) Water Loss Management in Distribution Systems: An Overview, SWITCH / Loughborough University, presentation at Water Management Training, 18–19 May 2008, Alexandria, Egypt

[11] IB-NET (2011) IB-NET database, www.ib-net.org accessed 12th June 2011

OECD (1999) *The Price of Water: Trends in OECD countries*, OECD, Paris, France

OECD (2007) *Financing Water Supply and Sanitation in ECCA Countries and Progress in Achieving the Water-Related MDGs*, OECD, Paris, France

[12] Staddon C (2010) *Do Water Meters Reduce Domestic Consumption? A Summary of Available Literature*, Defra, London, UK

Dovey W J & Rogers D V (1993) The Effect of Leakage Control and Domestic Metering on Water Consumption in the Isle of Wight, *Water and Environment Journal*, 7 (2), 156–160

[13] Thames Water (2011) 2011 June Return, http://www.thameswater.co.uk/cps/rde/xbcr/corp/thames-water-june-return-tables-2011.pdf accessed 17th June 2011

[14] Rogers P, de-Silva R & Bhatia R (2002) Water is an Economic Good: How to Use Prices to Promote Equity, Efficiency and Sustainability, *Water Policy*, 4, 1, 1–17

[15] Rogers P (2011) Water Governance: the Relevance of Price Policy, *The Water Leader*, 3, 34–35, June 2011, Institute of Water Policy, NUS, Singapore

[16] World Bank (2002) *Water Tariffs and Subsidies in South Asia,* World Bank, Washington DC, USA

[17] www.iandp.net

[18] http://www.abb.co.uk/cawp/seitp202/869efcdcddeba8dfc 12574da0050c517.aspx accessed 12th July 2011

[19] www.senceive.com

[20] Liemberger R (2008) The non-revenue water challenge in low and middle income countries, *Water* 21, June 2008, p 48–50

21 WHO / UNICEF (2002) *Global Water Supply and Sanitation Assessment 2000*, WHO, Geneva, Switzerland

22 Frauendorfer R & Liemburger R (2010) *The Issues and Challenges of Reducing Non-Revenue Water*, ADB, Manila, Philippines

23 Ofwat (2010) A drain on society – what can be done about water debt?, Ofwat, Birmingham, UK
Oxera (2006) Watered-down incentives? Bad debt in the water industry, Agenda, March 2006, Oxera, Oxford, UK

24 http://www.ib-net.org accessed 16th June 2011. Data is for 2009.

25 Al-Musallam L B A (2006) Urban Water Sector Restructuring in Saudi Arabia, Presentation to the GWI Conference, Barcelona, Spain

26 Ardakanian R & Martin-Bordes J (2009) Proceedings of International Workshop on Drinking Water Loss Reduction, UN Campus, Bonn, 3–5th September 2008, UNW-DPC, Bonn, Germany

27 Muhairwe W T (2009) Fostering Improved Performance through Internal Contractualisation, Presentation to the 5th World Water Forum, Istanbul, Turkey, March 2009

28 NSWC (2010) National Water and Sewerage Corporation, Annual report 2009–2010, NSWC, Kampala, Uganda

29 http://en/breivoll.no

30 Sydney Water (2010), Water Conservation Strategy, 2010–15, Sydney Water, Sydney, Australia

31 www.i2Owater.com

32 www.sanipor.com

33 Chartres C & Varma S (2010) *Out of Water: From Abundance to Scarcity and How to Solve the World's Water Problems*, FT Press, New Jersey, USA

34 Dworak T et al (2007) EU Water Saving Potential, EU Environmental Protection Directorate, Berlin, Germany

35 Environment Agency (2009) *Water for People and the Environment. Water Resources Strategy for England and Wales,* Environment Agency, London, UK

36 Moskvitch K (2011) Internet of things blurs the line between bits and atoms, BBC News 3rd June 2011

37 *Financial Times* (2011) Water risk database backed by US groups, FT, 16th August 2011, http://www.ft.com/cms/s/0/ac39db6c-c75a-11e0-9cac-00144feabdc0.html#axzz1VEbLWj8H, accessed 16th August 2011

38 Green C (2010) Presentation to CIWEM Surface Management Conference, SOAS 2nd June 2010

39 Kurlansky M (2002) *Salt: a world history*, Bloomsbury Publishing USA, New York, USA

40 Earth Institute (2006) It's 2025. Where Do Most People Live? The Earth Institute at Columbia University, 18 Jul. 2006. Web. 1 Jun. 2011.

41 Adams F D (1954) *The Birth and Development of the Geological Sciences*, Dover, New York, USA

[42] World Bank (2004) Seawater and Brackish Water Desalination in the Middle East, North Africa and Central Asia: A Review of Key issues and Experience in Six Countries, The World Bank, Washington DC, USA
WWF (2007) *Making Water, Desalination: option or distraction for a thirsty world?*, WWF International, Gland, Switzerland

[43] Examples of companies active here include:
Forward Osmosis: Modern Water (UK) www.modernwater.co.uk
Aquaporins: Aquaporin (Denmark) www.aquaporin.dk
Membrane-condensation & waste heat: Terra Water (Germany) www.terrawater.de
Carbon nanotubes; Still on the lab bench – New Jersey Institute of Technology (USA)
Energy recovery: Energy Recovery Inc (USA) www.energyrecovery.com

[44] GWI (2011) Low-energy membrane bid set to win UAE desal contract, GWI Briefing, 2nd June 2011

[45] GWI (2011) Is thermal desalination greener than it looks?, GWI Briefing, 24th February 2011

[46] Hurlimann A & Dolnicar S (2010) When Public Opposition Defeats Alternative Water Projects – the Case of Toowoomba Australia, *Water Research*, 44 (1), 287–297

[47] http://www.blogtoowoomba.com/entry.php?w= toowoombawatervote&e_id=565

[48] Chartres C & Varma S (2010) *Out of Water: From Abundance to Scarcity and How to Solve the World's Water Problems*, FT Press, New Jersey, USA

[49] GlobalData (2011) Power Generation from Wastewater Treatment Sludge – Power Generation, Sludge Management, Regulations and Key Country Analysis to 2020

[50] Schwarzenbeck N, Pfeiffer W & Bomball E (2008) Can a wastewater treatment plant be a powerplant? A case study, *Water Science and Technology*, 2008, 57 (10) 1555–61

[51] Ieropoulos I, Greenman J & Melhuish C (2011) Urine utilisation by microbial fuel cells; energy fuel for the future. Phys. Chem. Chem. Phys., 2012, Advance Article DOI: 10.1039/C1CP23213D

[52] www.algaewheel.com for a land-based wastewater treatment application and Howell K (2009) NASA Aims for Future Fuel from Algae-Filled Bags of Sewage: Can the aviation fuel of the future be grown in plastic bags of wastewater?, *Scientific American*, 12th May, 2009.

[53] Heidrich E S, Curtis T P & Dolfing J (2011) Determination of the internal chemical energy of wastewater. Environment Science & Technology, 2011 Jan 15; 45(2): 827–32

[54] Defra (2008) *Future Water: The Government's water strategy for England*, Defra, London, UK

[55] Rothausen S G S A & Conway D (2011) Greenhouse-gas emissions

from energy use in the water sector, *Nature Climate Change* 1, 210–219

[56] Neave G (2009) Advanced Anaerobic Digestion: More Gas from Sewage Sludge, *Renewable Energy World*, April 2009.

[57] Ofwat (2010) Financial performance and expenditure of the water companies in England and Wales 2009–10, Ofwat, Birmingham, UK

[58] www.encosltd.com

[59] Winpenny J & Heinz I (2010) *The wealth of waste: The economics of wastewater use in agriculture*, FAO Water Report 35, Rome Italy

[60] Labre J (2009) Key success factors for water reuse projects, Presentation at the 5th World Water Forum, Istanbul, March 2009

[61] GE (2008) The Water Energy Nexus, GE presentation to investors, 4th December 2008

[62] Bufler R & Salazar M (2011) Full cost recovery in water reuse schemes – examples and lessons learned from development cooperation projects. Poster presentation at the Governance and Finance – Water pricing policies session, Singapore International Water Week, Singapore, July 2011

[63] www.biomatrixwater.com

[64] Chartres C & Varma S (2010) *Out of Water: From Abundance to Scarcity and How to Solve the World's Water Problems,* FT Press, New Jersey, USA

[65] DEFRA (2007) Regulatory impact assessment – sewage collection and treatment for London, DEFRA, London, UK

[66] Bowen R (2010) Engineering and the idea of justice, 3rd Oxford Water and Membranes Research Event September 12–14, 2010, Oxford, UK

[67] Sen A (2009) *The Idea of Justice*, Allen Lane, London, UK

[68] Javier M (2011) The State of Global water Innovation, Cleantech Group LLC, London, UK

[69] Day R (2011) Ceantech Venture Capital in 2015, cleantechvc blog, February 2011, http://www.greentechmedia.com/cleantech-investing/post/cleantech-venture-capital-in-2015/ accessed 30th June 2011

[70] Norton Rose (2010) *Cleantech investment and private equity: An industry survey*, Norton Rose LLP, London, UK

[71] Boniolo M R (2011) Looking for Unconventional Solutions to Address Global Freshwater Issues, The Water Leader, 3, June 20-11, Institute of Water Policy, NUS, Singapore

[72] Tao G et al (2010) Energy reduction and optimization in membrane bioreactor systems, *Water Practice & Technology*, 5

[73] http://www.bluewaterbio.com/technology-hybacs-system.asp Note: The author is an advisor to the company.

Chapter 6

Affordability and Customer Choice

Water and wastewater management are characterised by the choices utilities make in developing and delivering their services, the choices made by politicians in allowing these services to be expanded and by policy and law makers when they consider what level of water and wastewater connection and treatment to adopt. It is evident that for the customer, choices are simple, yet more nuanced. They are typically expected to put up with what governments and municipalities think are good for them (if they think about them at all, let alone as customers) and yet, outside the constraints of traditional municipal utility service provision, they are being introduced to new worlds of freely available information and discretionary spending, worlds where people are able to exercise their choice about using utility services and what level of service they would like.

What happens when people don't have such a choice when it comes to access to safe water? For the urban poor, there is the trek to a standpipe or costly deliveries by a water carrier. Other choices have their own costs. There may be a private well, which will probably be unlined and thus susceptible to contamination as excreta from nearby latrines and leaking pits and sewers seep into the groundwater. Rainwater may provide a sporadic resource, if you have access to enough roof space and can afford the storage tanks. Such water does not stay fresh for very long. Urban sanitation when living without choice offers an equally risky lifestyle. Are people in developing countries supposed to be content to live in degrading conditions, presumably in the hope of a better beyond? Looking at the tone of some debates about service extension, some western anti-capitalist NGOs seem to assume that the complacent Orientalist myth of 'what will be, shall be' will forever remain the case, with ideological purity mattering more than human dignity. That results in policymakers and politicians blithely assuming that people are either unable – or unwilling – to make sentient choices when it comes to the provision of utility services.

One of the most corrosive myths surrounding water management is that people do not want to pay for adequate water and sanitation. Either they won't pay or they can't pay. In chapter seven we will see that the bounteous market for bottled water worldwide shows how people can and do pay for goods and services that they value. Looking at the surge in mobile telecoms in Africa and Asia provides valuable lessons in the development of utility services and affordability issues in the context of customer choice. When utilities grasp this, and are given the political and financial support, the scale, nature and trajectory of service extension can be transformed.

The assumption that people in developing economies are not able or willing to pay for sustainable water or sanitation services needs to be challenged as this is one of the principal impediments to universal access.

India – Why mobile telecoms markets move faster

Back in the mid 1980s, mobile phones were expensive, heavy and served by patchy networks. When I first visited Nokia's HQ in Espoo, near Helsinki early in 1987, much of the company's business remained focussed on its tradition of making lavatory paper and rubber boots, but Scandinavians were early adopters of mobile technology, and Nokia was Scandinavia's early adoptee. Mobiles were seen as the working tools of the developed world, the preserve of professional elites such as vets, financiers and politicians, something of possible interest to one in ten or twenty people in ten or twenty years time. Today, from Accra to Zunyi they are a part of everyday life, distinguished by their social ubiquity. Why is it that mobile phones, once seen as a luxury, have attained such a universal status, while access to safe water and sanitation, which ought to be universal, remains a luxury in developing countries?

Zafar Adeel, the United Nations University's Institute of Water, Environment and Health Director made a plea for increased investment in sanitation in April 2010: 'This simple measure could do more to...help pull India and other countries...out of poverty than any alternative investment.'[1] He noted that while in India 366 million people in 2008 had access to improved sanitation, the country now had 545 million mobile phone subscribers. He concluded: 'The tragic irony...India, a country now wealthy enough that roughly half of the people own phones, about half cannot afford the basic necessity and dignity of a toilet'. This is a telling point, yet the story that lies behind these figures is more compelling still.

As Dr Adeel observed, the expansion of mobiles in India has been remarkable. At the start of 2008, there were 234 million mobile phone subscribers, which had moved on to 752 million by the end of 2010. When a market more than triples in size within three years something dramatic is happening.[2]

As we have seen, access to safe water and sanitation tends to defy a strict definition. The World Health Organization classifies it as living within 1 Km of a safe water source for water (providing 20 litres per person per day) and with access to a private lavatory or latrine with a cover for sanitation. In India, it is interpreted as getting 40 litres a day from one public tap per 30 households or, on average, 162 people.[3] Household access to piped water is a clearer indicator. The definitions of improved sanitation also miss the point. Somebody can have a poor in-house system (a covered pit, for example) which is classified as 'improved', while a fully plumbed and serviced community block with separate entrances for men and women, replete with hand washing facilities, counts as shared and therefore along with 'unimproved' facilities, missing the target.

The United Nations' Joint Monitoring Project (JMP) has sought to monitor how water and sanitation access is developing, with reports published biennially since 2004. In India, as elsewhere, these assessments are a precarious process as countries change their criteria and the JMP interrogates their findings, so India's progress for water was revised down in the 2006 assessment and for sanitation in the 2008 assessment. Using the latest data, in 2000 208 million people in India had water piped to their homes and 261 million had improved sanitation. By 2010 260 million had piped water and 417 million improved sanitation. But these good works hit a demographic crunch. Population growth restricted the rise in percentage terms to 20% to 23% for water and 25% to 34% for sanitation, while the shift towards urbanisation meant the proportion served in urban areas fell from 49% to 48% for water, while nudging ahead from 52% to 58% for sanitation.[4]

Behind the sanitation figures lie sewerage networks that have not been expanded to cope with growing cities. New Delhi's sewers were designed to deal with three million people, not 14 million. A government survey in 2006 found that 28% of urban dwellers had household sewerage connections, with connection rates of between 25% and 30% from 1993 to 2006.[5] Just 3% of India's urban wastewater is properly treated, the rest being discharged into rivers, lagoons and the sea. According to the 2001 Census of India, eight out of the country's 3,119 towns and cities had complete sewerage and sewage treatment systems with 20% of towns and cities enjoying partial access, mainly in the larger cities. In 2003–04, 29% of the sewage generated in the largest (Class I) cities, covering 187 million people was treated, along with 4% in the Class II cities (43 million people). The absence of sewage networks means that 700,000 people make a living as 'scavengers', collecting and disposing of buckets of household excreta.[6] Then there is the challenge of slums.[7] The people exist, they are a major part of many cities' economy, but they are not part of the city of the politicians and planners. They rarely feature in utility projects or city statistics, and lacking tenant rights, they have no formal right to invest in formal water and sanitation services.

At a time when India's urban water and sanitation services are struggling to meet additional demand, let alone connect the unconnected, mobile phone services have moved from being a scarce luxury to the norm.

Uncrossing India's wires

The story of India's telecoms boom is one of choices and attitudes. Telecommunications formed part of a statist society, and access was traditionally allocated according to policy rather than practicality. As in other areas of Indian public life, this has been reformed over the past two decades, and since 1997 the Telecom Regulatory Authority of India has overseen the liberalisation and expansion of India's telecoms, supported by a pretty swift and robust data-gathering system. This was desperately needed, since until 1985 telecoms services were not seen as a national need and in 1990 there were just 4.6 million fixed line subscribers and a waiting list of 1.7 million. Throughout the 1990s, resources were put into telecoms with competitive mobile communications services being available from 1996. By 2000 there were 2.0 million mobile subscribers and 26.5 million fixed wire subscribers. Yet today, having peaked at 42 million lines in 2004, the fixed line service had eased back to 36 million by the end of 2010 (still with a waiting list), while there are 63 mobile phones per 100 people.[8] Meanwhile 51% of all people in India in 2010 had to defecate in the open.

While fixed wire remains mainly under public control, the opposite has happened for mobile communications. There were 41.4 million fixed line subscribers in March 2006, 82% shared by the two state-held carriers, compared with their 22% share of the 90.1 million mobile telecom subscribers.[9] By the end of 2010, the state-held carriers accounted for 83% of the fixed wire subscribers, but just 12% of the mobile subscribers. It is the new players who have changed the game and it is evident that their entry has been a popular one.

Why haven't the State-owned companies reaped the benefits of the mobile phone boom? I suspect it is because of their culture of providing services where they see fit rather than responding to demand. This has a parallel with the supply and demand led approaches towards water management. Another cultural difference is that the point of a private sector company is to thrive and developing new businesses and markets is central to meeting that need. State-controlled business can thrive and innovate, but this is not necessarily the reason why it exists.

India has liberalised its mobile communication markets so that people can chose the service they wish, rather than having to influence people in order to qualify for a traditional land line connection. Funding has grown with customer demand and the contrast in subscriber numbers tells its story.

Focussing in on India's water utilities

Financial and performance data on India's water is inconsistent, but number crunching a survey of twenty major utilities carried out by the Asian Development Bank in 2007 reveals some useful pointers. Five of these utilities were unable to provide any information on water losses, two on how much their subscribers consume and one on its spending.[10] As municipal entities, they do not appear to know exactly what services they are delivering, which is hardly encouraging. This looks like a classic case of municipal utilities that are allowed to exist without oversight because they appear to have lost their sense of purpose other than to exist and to provide employment and spheres of influence and opportunity for politicians and suppliers.

The utilities surveyed by the ADB provided water to 44 million people in 2006, 37 million via household connections and 7 million via public taps. Over the previous five years, they spent Rs 39.3 billion on their networks, or an average of Rs 2,014 per connection each year. However, nearly half the money came from one utility (Chennai) which served one tenth of the total people and is undergoing a major private-public capital spending programme, so average investment was appreciably lower. During 2005–06, 1.3 million people enjoyed new connections to water services, a 3% increase over the year.[11] This figure is encouraging, but in isolation, it is just an encouraging snapshot.

Just what has changed to make mobiles ubiquitous? It appears to be a fall in costs and a rise in opportunities. Tariffs progressively fell from Rs 15.3 per minute in 1999 to Rs 3.1 in 2002 and from then on, down to Rs 1.0 by 2006. Competition was stepped up with the third and fourth network licences being awarded in 2003 and licensing systems and roaming charges liberalised from 2005. Meanwhile, research by Vodafone comparing the various states in India has revealed a causal relationship between mobile phone penetration and economic growth; each 10% increase in mobile telecoms penetration boosts the economy by 1.2%. This boils down to efficiency, with farmers now sending photos of crops to experts to ensure they avoid excess pesticide use, along with access to market price data for crops and weather forecasts for timing the harvest and improving labour mobility.[12]

The fixed line telephone traditionally connected the civil servants, businessmen, politicians and other people who mattered at home and at work. For the first 46 years following India's independence in 1949, there was less than one line per 100 people, since there was little need for rapid communications under the 'Licence Raj' culture of a state-regulated economy. The long waiting lists for a new line were part of that culture, requiring bribes and connections to circumvent them. The crucial difference between fixed wire and mobile telecoms is that replicating networks is far easier without wires, especially at a low subscriber density. This means that the mobile market is better suited for competition, and it is clear that India is a competitive market, with 87.9% of the market shared by six national operators who are active in every state with the rest held by nine local and quasi-national players.[13] In this case, water and sanitation share the infrastructure challenges faced by fixed wire services.

What about funding? According to the World Bank, $34.6 billion was spent on telecommunications projects in India by the private sector between 2004 and 2008. The private sector accounts for 88% of mobile telecoms activity and the mobile telecoms sector accounts for almost all of India's private sector investment, so this is pretty a useful indicator. This works out at $110 for each subscriber gained during those years or $33 for each subscriber served during 2008, when $9.9 billion was invested as the network capacity was cranked up and the average subscriber base was 302.5 million. Global Water Intelligence estimates that spending on India's municipal water services was $1.32 billion in 2008, with a further $0.46 billion on wastewater and $0.65 billion on industrial water, or a distinctly diluted $3.85 on water and $0.80 on sanitation for each urban dweller. In the big cities surveyed by the ADB, the picture is the same, with Rs 182 spent each year on both services per person between 2002 and 2006 ($4.14).

Is this a fair comparison? Water and sanitation are meant to be growth sectors just like mobile telecoms, since the government had set a target for universal access by 2007, the 60th anniversary of Independence. Missing this target has its ironies, as Gandhi observed that 'sanitation is more important than independence'.[14] Funding is available for mobile telecoms because there was an attractively predictable prospect of a predictably attractive return on the investment. For water and sanitation, there is political stagnation. What little money there is does not always go where it should. As with fixed wire telecommunications, bribes are often needed to install new water connections, suggesting a reluctance to prioritise service extension. Bribery also extends to avoiding bill payments, for installing or repairing water meters and reconnecting water services.[15]

As far as water and sanitation goes, it is often felt that there are too few compelling drivers for a utility to expand their services. The government at the national and state level assumes that sanitation is expensive and unaffordable and that people don't want to spend their money on it, perhaps also because of a feeling that people from other classes or castes do not need these services. Likewise, it is assumed that there is no access to finance. Yet all the evidence contradicts this. If people can choose to spend on mobile phones, they also have the right to be able to make choices about spending on water and sanitation services which are currently denied to them. Poverty is a massive challenge in urban India, but people understand the benefits of safe water supplies and often have to pay more for unsafe vended supplies. Universal urban access is attainable because spending on municipal water and sewerage infrastructure is in fact a low-risk investment that provides exceptional social as well as economic returns.

This is the down-the-pan side of Indian municipal politics. Between 1992–93 and 2005–06, the proportion of the richest quintile of Indian households with flush lavatories rose from 94% to 98%, while rising from 14% to 34% for the middle quintile and 1% to 5% for the poorest. For the well-off, this is evidently a minor concern, especially since while India subsidises water and sanitation provision by $1.1 billion each year, just 20–30% of these subsidies reach the poor, as they do not usually enjoy these household services.[16]

In 1996, I met the Managing Director of the

Hyderabad Metropolitan Water and Sewerage Board. When I asked about the possibility of 24-hour water supplies, he assured me there was no need. As long as people know when they are getting their two hours of water each day, they can plan for it and are satisfied. This supply-led philosophy has some failings in practice. With the exception of Jamshedpur where the water utility is privately owned and operated by Tata,[17] no city in India offers 24-hour water services to more than a tiny minority of its people. Indeed, average water supply in the twenty utilities surveyed by the ADB was just 4.3 hours per day, with one utility providing water for 12 hours a day and two for less than an hour. It matters, because it affects meter reading and leakage control. While wealthier households can afford to have individual water tankers to provide continual household supplies, these are emptied every day prior to being refilled – as you want your water to be fresh – and this results in overall water usage at Western levels, despite their utilities providing a far lower utility. In Delhi, the average connected household spends Rs 2,000 every year dealing with this intermittent supply, 5.5 times as much as they spend on the supply itself. For the less well-off, it means household activities revolve around your allocated slot. Ten years on, Hyderabad still supplied a steady two hours of water each day.[18]

Asit Biswas is a Singapore based professor of water policy. His sister who lives in West Bengal has a fairly typical middle class experience with her local water utility. It provides, on a regular basis, two hours of water in the morning and one in the evening. It is stored in a tank so that her family has 24-hour availability. The water is not regarded as being fit to drink, so like many others, she uses a point of use household filter, which you buy on a three-year contract and needs servicing every six months. More money goes on pumping the water to the tank, which in turn needs to be cleaned every two months because it gets coated in algae and sludge. When Professor Biswas suggested to the utility that they could provide a 24-hour potable water service for 40% of this cost, they explained that they were not allowed to do so. This would interfere with its other role as a provider of jobs as patronage for local politicians. Any attempt at full cost recovery would also be unacceptable to politicians, so people end up paying more as they create a series of coping mechanisms.

In contrast to mobile phones, most urban water and sanitation services are controlled by municipalities who are unwilling or unable to develop their networks or improve service quality, except for the better off. While the potential for funding exists, there is a genuine scarcity of management control and political will.

Funding and families

Connection fees for water in Hyderabad are Rs 3,000–10,000 against a median household income of Rs 4,000 per month. While monthly interest rates of 1.25 to 2.50% for 12 to 24 month loans are on the high side by western standards, this is low when compared to microfinance loans, where interest rates are becoming a serious concern. So, the challenge here is to make finance affordable and manageable so that reasonable aspirations can be reasonably met. Families who manage their money on a day-to-day basis find it difficult to deal with monthly utility bills, let alone connection charges. A survey looked at households in a 'notified' slum in Hyderabad where 40% have mobile phones and 93% have electricity, 43% either got water from their own or neighbours' taps and 58% had individual latrines.[19] Nearly half of the latter households had suffered from lavatory blockages at least once in the previous month. 61% of those surveyed said their family's health would benefit from more formal services, 33% would save time by not having to collect water and 15% believed their social standing would improve. 60% of the households said they would be interested in such a loan either for water, mains sewerage or both.

Matters are not perfect in mobiles either. The increasing usage density in urban areas has yet to be matched in rural areas, which creates a digital divide when it comes to access to information. At the end of 2010 there were 141 telephones per 100 urban inhabitants against 30 per 100 rural inhabitants. This intriguingly high urban penetration is based on the number of SIM cards in circulation, and probably includes quite a few pay-as-you-go handsets which are not actively being used. That may well change, as according to iSuppli, an Indian market consultancy, mobile phone penetration could reach 97% by the end of 2014. Relating this to India's rural realities is probably unrealistic, but it remains evident that mobile phone subscriber growth and market penetration will outpace improved access to water and sanitation for quite some time.[20]

Yet initiatives can work in India's water and sanitation, as some recent examples show. Pune initiated a citywide sanitation programme in 1999 that has provided communal facilities for half a million people on time and on budget by insisting that only community and not for profit organisations were allowed to bid for the design, construction and maintenance of these units cutting down the scope for political cronyism. Too often, it is the will and the imagination that appears to be bankrupt.[21]

In Tamil Nadu, a sanitation connection project was developed on the basis of public support and partial subsidy funding.[22] 95% of people in Alandur in Tamil Nadu have household lavatories which discharged into open drains. A 'willingness to pay' survey found 97% of residents wanted a sewerage and sewage treatment system and would pay up to Rs 2,000 per connection. The project cost Rs 340 million for the 146,000 residents, who contributed towards 23% of the project's funding. In both Alandur and Hyderabad, it was the prospect of access to an improved service and an affordable manner of paying for the connection that is making service extension possible under these initiatives.

Two pilot projects offer another approach, adopting what Karen Bakker refers to as 'elite archipelagos' of water supply.[23] In Ahmedabad, metered customers in pilot projects will get a continual supply of 140 litres per person per day, with extra consumption being charged for.[24] Otherwise, water charges come as part of a property tax and there is a separate wastewater charge. A more upmarket version is being developed in Hyderabad for the specific supply of drinking water to housing colonies that are not covered by the utility.[25] Currently there are about a thousand housing colonies on the outskirts of the city that the Water and Sanitation Board does not supply. Groundwater is available, but it has very high fluoride levels. So the utility has selected eight housing colonies where a local reverse osmosis unit will provide purified groundwater for drinking at R150 per M^3 ($3.00).

Tata director R Gopalakrishnan believes that mobile communications will bring water and sewerage services in their wake as people newly armed with access to information and the means of communicating it are challenging government officials about their inaction: 'I think there are very, very dramatic changes happening.'[26] Patrick French observes that India is finally starting to 'come out' about prejudice and the poor.[27] Not surprisingly, the country's social differentials trickle down to access differentials when it comes to water and sanitation. It is no good any more saying that such unjustifiable differentials are an imperialist legacy or western idealism; they have become an unsustainable actuality which has to be confronted. Actualities also mean that optimism has to be constantly tempered. Preparations for the 2010 Commonwealth Games in Delhi and the ensuing fallout over cronyism and corruption demonstrate that (to use an old Indian Civil Service saying) sometimes, when 'you ask for increment, you get excrement'.

Fieldwork shows that even the worst off are ready and willing to pay an affordable amount for suitable water and sanitation. This is in contrast with the traditional assumption that the poor do not matter and there is no need to address their concerns. India's social and economic development depends on it becoming a connected nation, to water and sewerage as well as mobile telephony.

Kenya – can pay, will pay

Mobile returns outweigh their costs

A similar story has emerged in Kenya. Christopher Gasson runs Global Water Intelligence, perhaps the first company dedicated to providing the big picture on the costs and cost factors of water and wastewater worldwide. In 2010, he noticed the disparities between water and mobile phone connections and the costs of using and connecting to each service.[28]

At the start of 1999 Kenya had 5,300 mobile telecom subscribers, hardly surprising when it cost $149 to connect to the service and a monthly subscription of $16.57 even before a call was made.[29] By 2005, the market was taking off and there were 4.6 million subscribers, jumping to 16.3 million in 2008.[30] The spread is linked to a fall in the cost of handsets and the use of pre-paid SIM cards. The cheapest mobile phones dropped from KES 28,200 ($350) in 2004 to KES 4,460 ($55) in 2008 and Vodafone's Safaricom's cheapest handset in October 2010 was KES 999 ($12). There were 19.9 million SIM card users by 2010, equal to 51% of the population. While the market is dominated by Safaricom, the three other operators are competing hard for market share. The market is evolving in other ways, with 6.4 million connected to the Internet via mobile telecoms, in a country where there are 0.2 million fixed wire telephone lines.

It has become a serious business. Telecommunication service revenues as a percentage of GDP in Africa have grown from 2.1% in 1998 to 4.9% in 2004. In Europe and the Americas, the proportion varies between 2.8% and 3.3%. This reflects the intensity of capital spending as Kenya spent an average of $735 million per annum on private sector telecoms projects in 2004–08, when the number of subscribers rose from 2.46 million to 16.3 million, or $377 being spent per new subscriber. In 2008, $1.2 billion was spent, a year which saw 4.96 million new users at an investment of $242 per new user. Even in 2006, the competitive nature of the market was pushing its development. That year Kenya and Tanzania had 13.0 million mobile phone subscribers between them while Ethiopia had 1.5 million. At 76 million, the population of Kenya and Tanzania is the same as Ethiopia's but between them there are six operators while Ethiopia has a state monopoly.[32] Telecoms companies respond to opportunities, since a telecommunications network is characterised by relatively fixed operating costs, meaning that additional revenues generate higher margins, enables the regulator to push prices down further and allows further spending and service extension. So they seek to attract more customers and more call revenue in whatever way works best in gaining customer interest.[33]

This is a discretionary market, but its attraction was highlighted when ICT Africa carried out a survey of mobile telecom use in 2007–08. They found that subscribers were spending 16.7% of their monthly individual income and 52.5% of their monthly disposable income on mobile services. They do not see mobile phones as a luxury. Farmers use mobile phones to get access to micro-insurance, arrange cash transfers, meetings with their clients, transport their crops and ensure they have accurate information about commodity prices. Mobile internet services have moved data transmission from text messages to detailed reports.[34]

People are prepared to spend a lot of money on a service when they find that the benefits are clearly greater than the costs. At the same time, regulation can create a virtuous spiral of increased access, affordability, demand and funding.

Water and sanitation remain sessile

Having a mobile phone can be life-enhancing and transforming and so can having access to water and sanitation. Progress has been even more frustrating than in India and according to the UN's Joint Monitoring Programme, access to household piped water in urban areas actually fell from 49% to 44% between 2000 and 2008 partly due to the urban population leaping from 6.3 million to 8.5 million. Urban safe sanitation crept ahead from 26% to 27%, with marginal progress nationally, household piped water rising from 18% to 19% and safe sanitation from 29% to 31%. The urban household sewerage connection rate appears to have fallen from 42% in 1989 to 35% in 2003 and 34% by 2008.[35]

Access to water matters to people especially those who have limited access to politicians and policymakers. When the World Bank carried out a survey in Kenya's three largest cities in 2000, they found that 56% of respondents put water as their chief development priority against 11% for electricity and 9% for sanitation and sewerage. They paid on average KES 260 per M^3 ($3.22) for a mix of formal and informal services, wholly out of line with the average official tariff of KES 32 per M³ ($0.40). What people saw as a fair price for water (here KES 209–218 per M³, or $2.59–2.70) was not related to wealth, although poorer households had a lower threshold when it came to regarding water as expensive. As with paying for mobile telecoms, people will pay a fair price for affordable water services if utilities give them this option and inclusive financing mechanisms to cover connection costs can be provided along with any subsidy funding that is available being directed towards connection costs for the worst-off. The fair value of water is recognised irrespective of circumstances, but excessive tariff rises are quickly seen as unfair. What do people actually want? 93% of people living in Nairobi's Mashimoni slum depend on water kiosks, with 6% having shared connections and just 1% enjoying household access.[36] When asked about what they would like and would pay for, 7% opted for no change and 21% wanted improved kiosks, while 41% preferred improved yard facilities and 30% household connections. All those profiting from the status quo would have opted for no change, suggesting that a mighty majority are both willing and able to pay for a decent service.

The average piped household had a 1,825 litre storage tank costing KES 8,994 ($111) to smooth over their erratic water deliveries. The unconnected poor use 20-litre jerricans. The storage tank in effect becomes a connection charge due to the intermittent nature of the water supply. Again, service quality was a problem, with 50% of households being dissatisfied with their water and perhaps the richest 7–10% of the population is well served. Asking people what price they were prepared to pay for improved services was a case of stating the obvious. Yes, they were happy to pay less than they did now to vendors to get tap water either to the house or a communal yard.

A Water Act in 2002 has made a platform for reforming the sector, which was reorganised into seven regional water boards in 2005 along with a Water Services Regulatory Board which oversees tariffs and reporting. Matters may start to improve as the Water Services Regulatory Board has been producing annual assessments of the sector since 2008. They tend to be at the 'name and shame' stage, especially about poor reporting, but progress was being seen by 2008–09.

Impact: A Performance Report on Kenya's Water Services Sub-Sector by the Water Services Regulatory Board in 2010 gives an idea of some of the realities the sector faces. They note that while Nairobi Water estimated that 2.2 out of 3.2 million were served in 2008–09, in 2006–07, 1.1 million were directly covered and 0.6 million via water kiosks and communal water taps. There is little evidence of any great increase in piped connections since then. Nationally, tariffs had been unchanged for a decade to 2009 and are now slowly being revised. Since average revenues in 2008–09 only covered 98% of basic operating spending, they cannot usually maintain the networks, let alone any service extension or debt servicing. Meanwhile, much of the water produced does not appear to be used, with unaccounted-for water ranging from 7% to 86% among the larger providers, an average of 49%, while 37% of connections are classified as 'dormant' where water may be illegally used, and traditionally this was 'overlooked'. This means that the average national daily water consumption of 59 litres per person is almost met by the 57 litres of unaccounted-for water. Four out of the 23 larger utilities provide a continual water supply and the average of fifteen hours a day is appreciably better than in India, but continual supply is essential before public trust can be gained.

People hold their water services in low regard. The Kenyan chapter of the anti-corruption campaigners Transparency International found that in 2009 70% of households with piped water and 76% with non-household water treated it by boiling and disinfecting before use. With up to 10% of samples failing water quality tests in 2008–09, their concerns are well founded and treating unclean water further pushes up the cost of their water.[37]

Matters may be improving in Nairobi, as a Water Integrity Network survey in 2005 found 65% of respondents had a corruption-related experience in the past five years, while in 2009 the Transparency International survey found 12% of respondents knowing about corruption. Services converge, so in June 2010, the Nairobi Water Company made paying metered water bills simpler through Safaricom's mobile money transfer service M-PESA. Since then, four mobile operators have adopted this service across East Africa, ranging from single towns to national coverage in Uganda. Does it work? A team of Oxford-based researchers led by Tim Foster looked at M-PESA's performance when it was offered by the Kiamumbi Water Trust, a privately operated water town serving 620 households in a small Kenyan town.[38] Here, 99% of subscribers already had a mobile 'phone and between its introduction in December 2010 and July 2011, 74% of customers opted to pay their bills through M-PESA. For the customer, the cost of paying via M-PESA is half of a trip to the bank, while saving an average of 94 minutes in travel and queuing, for the utility, it lowers bill handling costs and gets payments in on time while all business is good business for a mobile operator. What it does not serve is those with mobiles who depend on a water kiosk. Could this become another driver towards universal access?

The Nairobi Water & Sewerage Company website promises 'a new beginning'. Although launched in 2009, it remained 'under construction' in 2010 before going on line in 2011. In 2007, 75% of people in Nairobi lived in poorer areas and various informal settlements that were not included in the municipality's water plans. The more the slums grow, the lower the proportion of people get access to safe water. In Kiberia, one of Africa's two largest slums, water is provided by vendors at $2.50 per M^3 against $0.46 per M^3 for the better off in Nairobi. A more hard-headed view is provided by the Kenya Alliance of Resident Associations who, in 2007, found that while Nairobi as a whole had 74% access to mains water, just 18% of the poor enjoyed household access against 86% of the non-poor. Indeed, in the informal settlements, 1% of households have direct water access and 44% of households shared taps. The other 56% spent 90–120 minutes a day obtaining their water. They also noted high rates of customer complaints and sewer blockages and blowbacks being commonplace.[39] With an average tariff of KES 40 per M^3, this works out as 14.6 M^3 each year or KES 861 ($10.7) per year, or 1.4% of GNI (Gross National Income) before other charges slip in. In Nairobi, the average household spends KES 913 per month (KES

476 by the poor), and each household spends on average KES 2,100 on a storage tank as well, to cope with the intermittent water supply. Compared with mobile phones, once connected to water it is eminently affordable.

According to Global Water Intelligence, capital spending for Kenya's 8.5 million urban dwellers on water in 2007 was $42 million ($4.94 per person) with $5 million going on sanitation ($0.59 per person) with rather more on operating the services at $86 million for water ($10.12 per person) and $10 million for sanitation ($1.18 per person). However, the Ministry of Water and Irrigation budgeted KES 18.7 billion ($243 million) for water and sanitation in 2008–09 and budget allocations have risen by 245% since 2004–05.[40]

A connection charge in Nairobi costs $15 along with a monthly meter rent $2.50. These and a tariff of $0.46 per M^3 are not prohibitive but outside its central area, the same utility has a water deposit (connection charge) of $187.50, which hurts when the gross national income (GNI) per capita is $770. According to IB-Net, the International Benchmarking Network for Water and Sanitation Utilities, water connection charges at four Kenyan utilities ranged from 9% to 33% of Gross National Income per capita in 2006 or 12% to 45% of household spending. That makes the cost of connection more of an impediment than buying and using a mobile phone.[41]

The benefits of access to mobile telecommunications (like safe water and sanitation) are undeniable. People do not generally choose to spend a significant part of their household income on such a service unless they firmly believe they will derive a suitable benefit from it. Given a chance, they will spend the same on water and sanitation.

Here again, mobile phones are starting to change matters. Sixty community health workers in Kenya's Kuria have been equipped with internet-enabled digital phones to collect, store and share data about sanitation facilities.[42] Free Google-based software provides them with a searchable database instead of spending time and money on using hard copies.

People maintain the status quo when they see little scope for change, yet in a nationwide survey carried out in 2000, 56% of people living in three Kenyan cities saw access to water as their most important utility priority along with 9% favouring sanitation.[43] While the poor paid KSh 7.6 per day for their water against KSh 10.4 by the non-poor, they spent 42 minutes a day gathering it against just 15 by the non-poor. Just 5% of poor households had piped water, meaning that they pay the equivalent of $2.70 per M^3 for a 20-litre jerrican at a kiosk or $8.40 per M^3 for home delivered water against an average tariff of $0.46 per M^3 for piped water.

Water is a natural monopoly. Replicating a water network would be absurdly expensive (the extant water and sewerage infrastructure serving England and Wales is estimated to be worth £4,000 per person). As a result, we cannot strictly compare mobile communications with water and wastewater, but they are both utilities and they highlight that where a utility is allowed to invest and maximise its customer base it enjoys the potential to boost human happiness, while driving down costs as it taps into layers of hitherto unsatisfied demand.

Evidence exists in Kenya – just as in India – that there is popular demand for access to water that people are ready and willing to pay for. Poor services are costly for almost everybody and those costs would be better spent on a suitable service.

Connecting the unconnected

India and Kenya illustrate two conflations of water and sanitation service stagnation at a time when mobile telecoms were taking wing. Yet people in other cities have seen their services expand in a way almost as unimaginable as the mobile telecoms boom. Cambodia offers a good example.

Phnom Penh – transforming service delivery

Phnom Penh's Water Supply Authority was corporatised in 1996. Between 1996 and 2001 it received $94 million of donor assistance and $56 million in loans from the ADB, the World Bank, Japan's JICA and the French Government. The transformation of the city's utility performance is one of the most dramatic on record. In 1993, there were 25,960 recorded customers, somehow broken down into 12,980 of which did not receive water and 13,901 'customers' who did receive water but were not registered. The distribution network was refurbished between 1994 and 1999 and the network expanded from 2000. From 1998, a programme to regain the public's trust in the utility and to encourage billing payments went into operation, including improved information availability and simplified payments. Tariffs were revised in 1997 and 2001 and have been unchanged since.

Service delivery, 1993–09	1993	1995	2000	2005	2009
Connections	26,881	28,654	67,016	138,226	192,514
Metered connections	3,391	15,023	66,905	138,226	192,514
Collection efficiency	48.0%	50.0%	99.5%	99.8%	99.9%
Non-revenue water	73.0%	64.0%	33.0%	9.0%	5.9%
Revenues (million Riels)	1,400	5,500	25,125	64,679	105,780
Gross profit (million Riels)	-700	1,300	4,191	23,380	33,313

Staff per 1,000 customer connections fell from 20 to 3.2, meaning the utility serves its customers rather than itself. Water quality now meets WHO standards, with production having increased from 65,000 M^3 per day to 300,000 M^3 per day. The coverage area has increased from 20% to 90% with pressure at 2.5 bars against 0.2 bar and 24 hours per day water delivery against ten hours per day. The reduction in non revenue water has resulted in a saving of 165,000 M^3 of water per day, equivalent to $100 million in water treatment facilities. Service extension has been mobilised through a 'water for all' programme which has connected 18,862 low income families, providing subsidies of $5 per family per month, through lowering the connection cost and lower tariffs. The utility took the needs of Phnom Penh's people seriously, and it has in turn gained their support and indeed admiration.

Public or private?

Private sector and privatisation has changed the mobile telecoms sector. While PSP, private sector participation, is not necessarily the cure all for water and sanitation, the fact that such an option exists has forced the nature of service provision to evolve. Utilities have been forced to consider the customer rather than the politician and to realise that the former will be their longer-term paymasters. Huge challenges remain, especially when it comes to sewerage. LYDEC serving Morocco's Casablanca is adopting an integrated water and sewerage approach which was the aim in Bolivia's La Paz and El Alto, where political and change ended the contract early in 2007. In Manila, sewerage extension has underperformed while water extension has outperformed. In Phnom Penh, sewerage and sewage treatment remains a distant prospect for the municipal utility and will require new sources of funding, some perhaps from the private sector. In every case the challenge lies in reconciling the more immediate rewards in connecting to household water services against the less tangible benefits of a more costly sewerage service.

This is not a conflict between the private sector and the state. It is a conflict between two attitudes, about where the customer belongs in a utility's priorities and how flexible the utility is prepared to be in extending its services. It is also clear that this a tougher task when it comes to sewerage.

The human factor's tangent in Tangier

The World Health Organization[44] believes that one of the chief benefits of improving access to water and sanitation is that it gives people more time to work. Assuming 20 billion extra working days could be generated, they calculated that this would lead to $63 billion in productivity gains. As with a lot of economics this was developed on assumptions about how people behave, so what happens when we factor in the human aspect?

A field trial started in 2007, where people living in 850 low income households in Tangier were offered interest-free loans for water connections repayable over three to seven years depending how far they lived from the water mains.[45] Customers paid the full cost of connections and water and the loans were subsidised. The work was carried out by Amendis (Veolia Environnement's Moroccan arm which serves the city) using various funds available for social projects, and the Massachusetts Institute of Technology's JPAL Poverty Lab developed a trial to examine its impact. 70% of those offered the subsidised loan took it up, even though this doubled their water bills when repaying the loan during the period at DH 15 ($2) per month. As the area already had street pumps which provided potable water, this was about comfort and convenience rather than public health. By joining the scheme, people would in effect be buying time and the sense of belonging that comes with having a household connection. While the subsidised loans were available to all in the trial, only some were fully informed about the loans and provided with support to deal with the paperwork necessary, since poorer households in old city centres need to have their status formalised.

As expected, people enjoyed significant gains in time. But instead of doing extra work, they spent it relaxing, shopping and going out. So, humans the world over are humans, not machines, and the subsidised loan for water connections created a significant rise in the happiness of households. Since the urban poor rarely get the most

271

enjoyable jobs, being freed to spend even more time tied to such drudgery may not be their idea of a globally connected utopia.

Other subtleties have been found in various field trials. People may, for example, use alternative water sources such as rainwater at certain seasons as this is cheaper than piped water. Likewise, as sanitation services become more advanced, the quantifiable benefits level off, while further benefits gained from sewerage and wastewater treatment such as the avoidance of environmental harm cannot so easily be monetised.[46]

Does this undermine the case for new investment? It does not, since other benefits come to light when theories are replaced by realities. People are all capable of making valid decisions and those decisions do not merely extend to choosing leisure over labour or exploiting seasonal water opportunities; they also mean that people are rational and capable of deciding what is an appropriate and affordable level of payment for a particular service.

Models and assumptions are useful for sketching out projects, but nothing beats on the ground experience. People are people, they want better access to water for many reasons and a desire for a better quality of life is not always expressed in terms of material gains.

Conclusions – new technologies and new attitudes

Water is a very good servant, but it is a cruel master.
Cecil Roberts, Adrift in America

People without access to telecommunications are leapfrogging traditional hard wire services for mobile networks. The ability to gain first-time access to a service through a disruptive technology has many implications for water and sanitation. It shows how ossified traditional assumptions and constraints do not apply when technologies, techniques and attitudes change. There is still the problem of water being a natural monopoly, which means that competition can never be used as a driver, except when artificially imposed.

One way of considering this is to consider whether those without adequate sanitation should leapfrog sanitation to sewerage. As we have seen earlier, the benefits of universal sewerage rather than partial connection mean there is a strong case for making sewerage extension schemes compulsory, especially when an overwhelming number of people choose to pay for this service and a programme is developed that makes their connection affordable. For water, the choice is simpler, as once the mains network is in place, additional costs for household connections can be carried out at the household level without being impacted by those in the area who for the time being choose not to be connected.

Globally, there is a debate emerging. Can other people choose what water they drink and should people be allowed to choose what water they drink? The same applies for their sanitation services. This leads to the question: Why is the public disconnected from these processes? This especially applies to slum dwellers, who should not be condemned to passively accept their fate. With information and communication, pressure for change can be created and these changes deserve to be taken up at the national level. Development ought to be about developing the potential for people's lives, rather than flattering the egos of the elites.

This also leads us to another pressing debate, the debate about bottled water. Why do people wish to pay more, much more for bottled water rather than tap water? Is it healthier or in any way better for you? Is it an environmentally friendly option? Should it be allowed at all? This debate deserves a chapter on its own, which follows.

---

[1] United Nations University (2010) UNU-INWEH Press Release, April 2010: Greater Access to Cell Phones than Toilets in India India has more mobile phones than toilets: UN report, *Daily Telegraph* 15th April 2010 http://www.telegraph.co.uk/news/worldnews/asia/india/7593567/India-has-more-mobile-phones-than-toilets-UN-report.html, accessed 27th September 2010

[2] Telecom Regulatory Authority of India Press Release No 16/2009 Telecom Regulatory Authority of India Press Release No 11/2011

[3] McKenzie D & Ray I (2009) Urban water supply in India: status, reform options and possible lessons, *Water Policy* Vol 11 No 4 pp 442–460

[4] WHO / UNICEF JMP (2004) *Meeting the MDG drinking water and sanitation target: A mid-term Assessment of Progress*, WHO, Geneva, Switzerland WHO / UNICEF JMP (2006) *Meeting the MDG drinking water and sanitation target*, WHO, Geneva, Switzerland WHO / UNICEF JMP (2008) *Progress on Drinking-water and Sanitation: special focus on Sanitation,* WHO, Geneva, Switzerland WHO / UNICEF JMP (2010) *Progress on Drinking-water and*

Sanitation 2010 update, WHO, Geneva, Switzerland

5   WHO / UNICEF (2010) Joint Monitoring Programme for Water
    Supply and Sanitation Estimates for the use of Improved Sanitation
    Facilities, India, updated March 2010,
    http://www.wssinfo.org/fileadmin/user_upload/resources/IND_san
    .pdf accessed 28th April 2011

6   Panse D B (2006) Ecological sanitation – A need of today! Progress
    of ecosan in India, GTW / DWA, Berlin, Germany

7   ADB (2007) Dignity, Disease, and Dollars: Asia's Urgent Sanitation
    Challenge, www.adb.org/water/operations/sanitation/pdf/dignity-
    disease-dollars.pdf.

8   Telecom Regulatory Authority of India (2007) A journey towards
    excellence in Telecommunications
    http://www.trai.gov.in/achievment.pdf
    Telecom Regulatory Authority of India Press Release No 11/2011
    Telecom Regulatory Authority of India 2005–06 Annual Report

9   Telecom Regulatory Authority of India Press Release No 1/2004
    Telecom Regulatory Authority of India Press Release No 11/2008
    Telecom Regulatory Authority of India Press Release No 11/2011
    Telecom Regulatory Authority of India Press Release No 08/2010
    Telecom Regulatory Authority of India 2005–06 Annual Report
    Telecom Regulatory Authority of India 2008–09 Annual Report
    Telecom Regulatory Authority of India (2010) The Indian Telecom
    Services Performance Indicators January–March 2010
    Telecom Regulatory Authority of India (2008) The Indian Telecom
    Services Performance Indicators January–March 2008

10 ADB (2007) 2007 *Benchmarking and Data Book of Water Utilities
    of India.* Ministry of Urban Development, Delhi / Asian
    Development Bank, Manila

11 See www.chennaidesal.info

12 Shina N (2007) Mobile Telecommunications and Development: the
    role of Public-Private Partnership in India, UK Development Studies
    Association "Information, Technology and Development" Study
    Group Mobiles and Development: infrastructure, poverty,
    enterprise and social development, Manchester, 16 May 2007
    Vodafone Group (2009) India: The Impact of Mobile Phones, The
    Policy Paper Series, Number 9, January 2009.

13 Telecom Regulatory Authority of India 2005–06 Annual Report
    Telecom Regulatory Authority of India Press Release No 11/2011
    Telecom Regulatory Authority of India 2008–09 Annual Report

14 Gandhi – quoted by Mammohan Singh, Prime Minister's speech, at
    the Third South Asian Conference on Sanitation, New Delhi, 18th
    November 2008

15 World Bank, Private Participation in Infrastructure Database:
    www.ppiaf.org
    GWM 2011 (2010) *Global Water Markets 2011*, Global Water
    Intelligence, Oxford, UK

[16] Foster V Pattanayak P & Prokopy L S (2003a) Do current water subsidies reach the poor? Water Tariffs and Subsidies in South Asia Paper 4, Water and Sanitation Program-South Asia.
Foster V Pattanayak P & Prokopy L S (2003b) Can subsidies be better targeted? Water Tariffs and Subsidies in South Asia Paper 5, Water and Sanitation Program-South Asia.

[17] http://www.juscoltd.com/

[18] ADB (2007) 2007 *Benchmarking and Data Book of Water Utilities of India*, Ministry of Urban Development, Delhi / Asian Development Bank, Manila
McIntosh, A C. (2003) *Asian Water Supplies: Reaching the Urban Poor,* Asian Development Bank and IWA Publishing: London.
ADB (2009) India's Sanitation for All: How to Make it Happen, ADB, Water for All 18, Asian Development Bank, Manila
Dueñas C (2008) Crusading for Human and Environmental Dignity. www.adb.org/Water/Champions/pathak.asp.
Tigno C (2008) Country Water Action: India, Toilet Technology for Human Dignity, Asian Development Bank, Manila
Dueñas C (2003) Water Champion: Almud Weitz – Breaking Barriers in Serving the Urban Poor, www.adb.org/water/champions/weitz.asp
McKenzie D & Ray I (2009) Urban water supply in India: status, reform options and possible lessons, *Water Policy* Vol 11 No 4 pp 442–460

[19] Davis J et al (2008) Improving access to water supply and sanitation in urban India: microfinance for water and sanitation infrastructure development, *Water Science & Technology*, doi: 102166/wst.2008.671

[20] Vodafone Group (2009) India: The Impact of Mobile Phones, The Policy Paper Series, Number 9, January 2009
iSuppli press release 22-09-2010

[21] ADB (2007) Dignity, Disease, and Dollars: Asia's Urgent Sanitation Challenge, www.adb.org/water/operations/sanitation/pdf/dignity-disease-dollars.pdf

[22] WSP-SA / MoUD (2008) Technology Options for Urban Sanitation in India, Water and Sanitation Program-South Asia, The World Bank, Delhi, India

[23] Bakker K (2010) *Privatizing Water: Governance Failure and the World's Urban Water Crisis*, Cornell University Press, New York, USA

[24] Ooska News (2011) Ahmedabad Authorities Launch Pilot Project for 24x7 Water Supply, Weekly Water Report Southern and eastern Asia, Ooska News, 28th June 2011, 5, 24

[25] Ooska News (2011) Hyderabad to Provide Purified Water – At a Price, Weekly Water Report Southern and eastern Asia, Ooska News, 28th June 2011, 5, 24

[26] India: Land of many cell phones, fewer toilets, *Bloomberg Businessweek*, 31st October 2010.

[27] French P (2011) *India: A Portrait*, Allen Lane, London, UK

[28] Global Water Intelligence (2010) Drinking from your mobile phone, Insight, 5th August 2010

[29] ITU (1999) World Telecommunication Development Report 1999

[30] http://www.balancingact-africa.com/news/en/issue-no-448/telecoms/700000-orange-mobile-subscribers-in-kenya-since-september

[31] ITU (2006) *World Telecommunication / ICT Development Report 2006 Measuring ICT for social and economic development*, ITU, Geneva, Switzerland

[32] http://www.ictworks.org/news/2010/08/27/3-reasons-why-kenyan-mobile-tariff-price-war-matters-ict-companies, accessed 12th October 2010
http://allafrica.com/stories/201003250883.html, accessed 12th October 2010

[33] Business Daily, Nairobi, 17th February 2010 (http://allafrica.com/stories/201002161110.html, accessed 11th October 2010)

[34] Gillward A, Milek A & Stork C (2010) Gender Assessment of ICT Access and Usage in Africa. Volume One 2010 Policy paper 5, Research ICT Africa, Cape Town, South Africa

[35] http://www.wssinfo.org/fileadmin/user_upload/resources/KEN_san.pdf

[36] Mwangi P (2011) Innovative Techniques and Approaches for Water and Sanitation Services Improvement in Nairobi's Informal Settlements. Presentation at the Urban Inequalities forum, Stockholm World Water week, Stockholm, Sweden, 24th August 2011

[37] Transparency International Kenya (2010) Water Governance Study 2009 Summary

[38] Foster T et al (2011) Adoption and implications of mobile money for urban water services in Kenya. Presentation at the Financing Urban Infrastructure Workshop, Stockholm World Water week, Stockholm, Sweden, 23rd August 2011

[39] WSRB (2008) Impact: A performance report on Kenya's water services subsector, Water Services Regulatory Board, Nairobi, Kenya
WSRB (2009) Impact: A performance report on Kenya's water services subsector, Water Services Regulatory Board, Nairobi, Kenya
WSRB (2010) Impact: A performance report on Kenya's water services subsector, Water Services Regulatory Board, Nairobi, Kenya

[40] Production factor costs in Kenya 2004 (2005) Export Processing Zones Authority, Nairobi, Kenya
GWM 2011 (2010) *Global Water Markets 2011*, Global Water Intelligence, Oxford, UK

[41] The Kenya Alliance of Resident Associations (2007) Citizen's Report Card on Urban Water, Sanitation and Solid Waste Services in Kenya Summary of results from Nairobi, KARA, Nairobi, Kenya

[42] IRC (2010) Mobile technology: cell phones, Google Apps help bring basic sanitation and more transparency in Kenya, 10th September 2010, irc.nl/page/64771

[43] Gulyani S, Talukdar D & Kariuki R M (2005) Water for the Urban Poor: Water Markets, Household Demand, and Service Preferences in Kenya, Water Supply and Sanitation Sector board Discussion Papers Series, No 5, The World bank, Washington DC, USA

[44] Prüss-Üstün A, Bos R, Gore F & Bartram J (2008) *Safer water, better health: costs, benefits and sustainability of interventions to protect and promote health*, World Health Organization, Geneva, Switzerland

Hutton G & Haller L (2004) *Evaluation of the costs and benefits of water and sanitation improvements at the global level,* World Health Organization, Geneva, Switzerland

[45] Devoto F et al (2009) Household Water Connections in Tangier, Morocco, Poverty Action Lab, MIT, USA. http://www.povertyactionlab.org/evaluation/household-water-connections-tangier-morocco, accessed 2nd February 2011

Tremolet S (2011) Is water connection the key to happiness?, 19th January 2011 blog entry on Tremolet.com

[46] Hutton G (2010) The Costs and Benefits of Water and Sanitation Services, Presentation to the OECD Expert Meeting on Water Economics and Financing, OECD, Paris, France, 17th March 2010

Chapter 7

Bottled Water – The Paradoxical Premium

A bounteous market for bottled water worldwide shows how people can and do pay for water which they value. Bottled water is about paying a dramatic premium in order to buy into something marketed to make it feel more fashionable, convenient, pleasant to drink or even more healthy than piped water. It usually represents the other extreme from vended and tankered water services, which provide water to people without access to potable water sources, typically for a significant premium over what is paid by those with household connections. It is also distinct from water coolers or dispensers, which either use bulk bottled water (usually in 18.9 litre / 5 US gallon containers) or piped water (often with additional filtration) for providing cooled drinking water in offices and public places and water fountains and drinking fountains which dispense tap water in public places.

A mixed message in a bottle

I spent a sweaty few hours in Jakarta in March 1996, visiting a hazardous waste management site and being given the low down on how certain local companies would have their chemical wastes 'swept away' during the annual floods. It was the grinding taxi journeys through Jakarta's gridlocks which were the real revelation. On one side, the brown, anoxic mass that is the Ciliwung River bubbled away and on the other, hawkers pedalled bottles of water which could have come from anywhere. They did a good trade. The place felt ripe for a revolution, which indeed occurred within a couple of years and ever since then, two water concession contracts have begun the slow task of reconnecting the city to safe water services. Safe sanitation remains another question altogether.

Why drink bottled water? This seemingly straightforward question is nuanced by a plethora of circumstances. At one end is a small selection of super-premium waters that are either bizarrely packaged, sourced from reassuringly faraway places or lay claim to various health benefits. One of the best brand names is 'Glace Rare Iceberg Water' which costs $12–15 per 70 cl bottle. Here my niggle is with the word 'rare' since there is 11,350 times more water in glaciers than in rivers at any one time. In the middle is a range of waters that offer possibilities of purity and better taste, along with those which are a convenient source for the thirsty. Finally, as in Jakarta, there is the health lottery of bottled water which is bought by those with no other choice.

A couple of decades ago, I heard that Badoit was in short supply and might stop being shipped to the UK altogether.

I like Badoit's peculiar 'alkaliney' taste, so I secured a few cases. By the time I had polished these off, I had bought a carbon filter and my Thames Water tasted just fine. I had lost the habit and indeed, since then, the London Ring Main entered service and Thames Water have adopted more advanced water treatment methods which have lowered chlorine levels to an echo. But it is a habit that holds dear to many. Sometimes the bottle may be carbon neutral or specially recycled for the client, but I cannot recall visiting a banker, broker, fund manager or lawyer without the bottled stuff being the only show in town. The only exceptions are those water companies who have ordered their own corporate carafes. Nobody complained.

Every brand tells a story about meeting needs. Some purport to offer 'safe' water where none is available while others offer status, fashion or taste. They all carry a cost.

Fear – from marketing to the media

We have already seen that there are plenty of cases where municipal water may or may not have unhealthy levels of chemicals ranging from chlorine to pharmaceuticals or is contaminated by pathogens. These stories and others are just what the bottled water industry feeds from.

Rebecca Hill, a blogger at the *Guardian*, recently highlighted a splendidly asinine tap water health scare.[1] Real Water™ (www.drinkrealwater.com) who, apart from trademarking the notion of 'real water', seeks to warn us that tap water is 'stripped of electrons' during the distribution process, meaning that water molecules are delivered to us in a clumped-up state otherwise unknown to science. It is difficult to imagine how this occurs outside the vivid imaginations of what may be called the bottled-water-world but there we are. They take tap water, treat it and somehow reintroduce the electrons, providing us with a classic value-added tap water. Their water claims to remove toxins so quickly from our bodies during workouts that extra nutrition may be needed. Gyms sizzling with freshly flung toxins do not strike me as particularly pleasant places to work out in, but it is little surprise to find that the company's parent company is based in Las Vegas. Various celebrity endorsements are splayed across the company's surreal water site which deserves cult status for its unique understanding of science. As the site states, they 'do not make any health claims because it is illegal to do so in the United States', but dear readers are invited to find them for themselves in the wild frontiers of the Internet.

Environmental lobbies as well as the mainstream media are also in part to blame for the fear factor, since much emphasis is placed on how bad tap water is as a campaigning point or simply because scare stories make good copy. From the Environmental Working Group in the USA ('Over 300 pollutants in U.S. Tap Water', December 2009) to the *Daily Mail* in the UK ('Chlorine in tap water "nearly doubles" the risk of birth defects', 31st May 2008) every time an environmental NGO and the mainstream media attack water utilities, they scare more people into buying bottled water, which is the equivalent of driving cars down pavements in order to encourage people to walk.

Bottled water is not cheap, it needs to be sold. When people are poorly informed about water and health, fear sells.

A bit of a distraction, albeit a rather big one

Bottled water ought to be nothing more than a distraction. After all, any desire to outlaw it outright smacks more of rancid Puritanism rather than any higher concern. Instead of being an amusing irrelevance, it reflects many of the sector's failings and challenges. I have no argument with bottled water per se. It is obviously better for you than many other soft drinks, and there is no doubt that people need to drink water. In the USA, the guideline is 3.7 litres per day for men and 2.7 litres per day for women, 0.5 to 1.0 litres already coming from the food you eat. My difficulty with bottled water is its relative cost, its environmental impact, what it is marketed as being and the fact that it competes for consumer spending which could be better used on our water and wastewater infrastructure.

Bottled water is a big business. According to Datamonitor's 2010 *Bottled Water Global Industry Guide*, in 2009, 135 billion litres of bottled water were sold, generating revenues of $79.8 billion or an average of $0.59 per litre. They forecast the market rising to a round $100 billion by 2014, with 167 billion litres sold. Evidently, this is an imprecise art, as Zenith International estimated that 206 billion litres were sold in 2007 and the market will be worth $65 billion in 2012, while Euromonitor sees it as a 105 billion litre global market in 2005 growing to 170 billion litres by 2014, with the market value jumping from $99 billion to $180 billion. Globally, the Beverage Marketing Corporation found Mexico to be the most intense consumer of bottled waters at 224 litres per person per year in 2008 compared with 108 in the USA and 30 worldwide.[2]

In Britain, data from Zenith International shows consumption growing from 20 million litres in 1976 to 2.09 billion litres in 2009 or 34 litres per head in a market worth £1.4 billion.[3] Despite the efforts of the multinationals to push branded bottled waters, natural mineral water accounted for 61% of the market, along with spring water at 27%, leaving the rest to account for the remaining 12%.

BBC London and Scotland spent £365,368 on bottled water for its water coolers in 2008, along with an unknown amount on bottled water at BBC events worldwide.[4] In contrast, BBC Wales spent £1,489, since its water coolers generally use mains water. Could that money be better spent on making programmes? In 2007, bottled water sales in the USA were worth $11 billion when there was a $22 billion gap between urban water spending needs and funds.

Tap water – aesthetics and deeper concerns

Tap water does have a perception problem. There are countries where people rightly perceive it to be either unattractive or unhealthy. The OECD conducted a survey of 10,000 households in ten member states in 2008. In the Netherlands, 95% of respondents were satisfied with their tap water, with 62% of those who were dissatisfied having taste concerns. In Australia, a 72% satisfaction rating saw 56% of those who were dissatisfied citing taste. At the other end, Korea and Mexico respectively endured 30% and 21% satisfaction ratings, with 87% and 92% respectively being concerned about health. The greater the dissatisfaction with tap water, the greater the health concern – a shift from aesthetics to actualities. Mexico's bottled water habit is evidently based on widely felt concerns.[5]

Schools – places to learn or places to earn?

Sometimes you wonder if other more cynical drivers exist when it comes to pushing vended drinks over tap water. Radha Sethi checked 198 schools in Manchester in 2000 and found that 62% did not have accessible taps, forcing children either to buy vended drinks or to go thirsty.[6] The World Health Organization believes that children should drink 3–4 glasses of water during the school day, but many in Britain have other priorities and the water fountains of our youth are being shut off. It may be a useful source of revenues, but it is also evident that forcing schoolchildren to buy sweetened drinks contributes to childhood obesity. There is also the irony of educational institutions letting children go thirsty, since classic symptoms of thirst are tiredness and lack of concentration. Oddly enough, it is the private water utilities who are trying to fill the gap. Kelda's 'cool school' scheme has donated £700,000 worth of water coolers to 250 primary schools in the poorer parts of Yorkshire, while Glas Cymru and the Welsh Assembly Government are carrying out a similar scheme in 100 deprived areas in Wales. My children were astonished when I told them that other children are denied access to water during the school day. They are still too young to fully understand the moral dereliction of some public administrators.

Is bottled water better?

It is reasonable to say that the perceived purity of various bottled waters is a pretty contentious issue. These examples will give you a flavour of the debate. Paul Hunter, when director of the Chester Public Health Laboratory, reviewed research on bacteria in bottled waters in 1992 and observed that 'all studies which have examined the microbiology of bottled mineral waters from retail outlets have found high total colony counts, particularly in still waters.' In Norway, Arve Misund's team sampled 56 bottled mineral waters which had been bought from across Europe in 1998 for 66 chemical elements. Just fifteen met all the European Union's drinking water quality parameters. In 2009 Sonish Azam and her colleagues at Montreal's C-crest laboratories found heterotrophic bacteria levels in over 70% of the bottled waters they sampled exceeded the United States Pharmacopeia's recommended limit of 500 colony forming units. The average microbial count for different bacteria in Canadian tap water is 170 colony forming units but Azam's team noted 'counts in some of the bottles were found to be in revolting figures of one hundred times more than the permitted limit'. A study of 35 bottled waters bought in Houston, Texas in 2007 found that four were significantly contaminated with bacteria.[7] This is a pretty small sample, in terms of bottles rather than brands, so the scope for statistical inferences is limited. When scientists examined the case of kidney stones in Parma in Italy, they found that stone formers 'did not follow a different dietary style from the rest of the population except for a high consumption of uncarbonated mineral water' or not less than two litres a day.[8]

Against this, Yale's Professor Stephen Edberg notes that there have been less than ten cases of bottled water-related illness in the USA across 35 years against the 16.4 million cases of gastro-intestinal illness noted by the US EPA in 2006. He also notes an average of one bottled water health recall a year between 1998 and 2008, tallying with Peter Gleick who noted twelve bottled water recalls in the USA between 1990 and 2002.[9] As we saw in chapter three, water utilities in the USA are suffering from decades of neglect. It may also be that bottled water quality is improving.

Alison Walker, writing in the *British Medical Journal* summed matters up tidily by stating 'the decision whether to trust tap water is ultimately left up to the consumer. But if fad or simply fancy advertising persuades you to buy bottled water, spare a thought for your purse.'[10] She wrote that in 1992, when our habits were rather more austere and tap water in England and Wales passed the Drinking Water Inspectorate's tests just 99.0% of the time, compared with 99.9% today.

The European Union's Health & Consumers' Directorate under its 'from the farm to the fork' policy, takes a close interest in water standards. Under EU rules developed between 1980 and 2009 mineral water 'may not be the subject of any treatment', excepting the removal of unstable compounds as long as this 'does not alter the composition of the water as regards the essential constituents which give it its properties and the addition or removal of free carbon dioxide by exclusively physical methods'. Likewise, spring water is defined as water intended for human consumption 'in its natural state and bottled at source', which can undergo a narrow range of approved treatments so that it complies with the EU drinking water directive. Basic testing is required every day (hundreds of times a day for the larger facilities) with a full analysis at least once a year. Waters imported from outside the EU also have to meet these standards.[11]

This means that in most of Europe, mineral water has to be bottled at source, naturally free from pollution, parasites and pathogens and that both mineral and spring water ought to be quite safe to drink and be properly labelled. Spring water has to come from an underground source and be bottled at source, but does not need to be officially recognised for its chemical characteristics, as long as it meets the health standards. It has to meet microbiological standards rather than be free of microorganisms and can be disinfected to remove them if needed. In both cases, this is a better state of affairs than implied in the USA and Canada above. Bottled water has to be recognised as what it is – value-added tap water. So, in Europe, a reasonable degree of consumer protection exists for the concerned consumer. Much of the rest of the world does not enjoy access to such information or protection.

While evidence that drinking bottled water is unhealthy is mixed and relates the way it is regulated, evidence to suggest that it is healthier than tap water is thin on the ground.

Is bottled water sustainable?

Somehow or other, drinking bottled water has been marketed as a sign of being green. Peter Gleick's team suggests that bottled water requires 5.6–10.2 mega joules

(MJ) per litre against 0.005 MJ per litre for tap water to get to the glass, a ratio similar to that of the cost of enjoying each beverage.[12]

Let's put that into context. If everybody in Britain was to switch to bottled water and drank 2 litres a day that would work out at 730 litres per person each year, or 4,088–7,466 MJ per annum. Nationally (assuming baby feeds would be made up this way as well) that works out at 245,280,000–449,760,000 GJ compared with Britain's total energy use of 9,850,000,000 GJ in 2008. So, bottled water would boost our energy consumption by 2.49%–4.57%, which is not an insignificant amount. This is a slightly provocative exercise, but it has a point especially when you consider how much bottled water per capita is already consumed in Mexico.

The energy cost can be reduced by moving from glass to plastic bottles (but don't forget the emerging issue of what water may or may not be getting from plastic packaging) and plastic bottles can also be made lighter. Bottles can also be recycled and some indeed are. In the USA, the PET bottled water recycling rate rose from 17% in 2004 to 31% in 2009.[13] Tap water needs no such packaging, which tends to make the comparison academic. Much of the debate is indeed academic, so Peter Gleick's *Bottled And Sold: The Story Behind Our Obsession With Bottled Water* is a useful starting point for going into more detail on bottled water and sustainability, with an industry viewpoint being put forward by the International Bottled Water Association (www.bottledwater.org) in a debate that is just starting to roll.[14]

For me, the idea of transporting a locally available ultra-low-cost commodity to another part of the world where it is equally available in terms of quality and quantity does in all honesty appear to be somewhat absurd. There are cases where resource imbalances make bulk water shipments a feasible approach and indeed humanitarian exceptions when a natural catastrophe or environmental or political circumstances prevent any reliable sources of water from being immediately mobilised. Even so, watching US soldiers trying to win 'hearts and minds' by handing out bottled water to people in Iraq whose water plant had been blown up was depressing at best. It might not have made glibly good TV, but sending out some mobile pumps, generators and filtration units would, as usual, have actually done the job better and faster.

By any reasonable criteria, bottled water represents a notably energy and resource intensive alternative to tap water.

Size does matter

There is an unspoken belief that by wearing running kit and squirting water down your gullet (preferably framed by a setting sun) you will lose weight and get fit. Indeed, sport bottles may well help through all the energy expended when trying to get a drink out of them as you wrestle through their excess packaging. It is odd how like is not usually allowed to be sold with like in different packaging (sports cap in 750 ml bottles and non-sports cap in 500 ml or litre bottles and so on) but occasional direct comparisons can be found. Twelve packs of half-litre bottles of Highland Spring cost 25–36% more with a sports cap, while there is a 42% premium for a sports cap when buying six packs of 500 ml from the Essential Waitrose range. Oddly enough, there is no price difference for the latter when you buy single bottles. This highlights how the cost of the water has little to do with the cost of the product. This gets dizzying when we look at a supermarket's range. These are for Sainsbury's Caledonian Scottish Water, another mid-range mineral water:[15]

Type	Unit	Pence per M^3
Still	6 x 500 ml	45,000
Still	1 x 5 litres	20,000
Still	4 x 2 litres	17,250
Still – sports cap	1 x 750 ml	90,667
Still – sports cap	6 x 300 ml	88,333
Still – sports cap	6 x 500 ml	45,000
Sparkling	1 x 330 ml	100,000
Sparkling	6 x 500 ml	55,000
Sparkling	4 x 2 litres	17,500

A 5.8 fold price range for what is the same product. Meanwhile, a two-litre bottle of Sainsbury's Basics still or sparkling table water (bottled water) costs 8,000 p per M^3. The 300 ml version is a 'Kids' size which is the current trend-setter, which is child's play given the premiums they can command. Yet the pricing can be just as perplexing. Water from a six pack of 'Kids' Highland Spring costs 32% more per unit than standards packs but 3% less when bought in twelve packs.[16] Perhaps I am a killjoy, but my children use refillable bottles with the same snazzy, fiddly sort of cap, with 'Dewis Dŵr' (Welsh for 'Choose Water') emblazoned on the side.

Looking across the supermarkets' labels and the traditional brands, it is interesting to note how often there is no difference between the prices of carbonated and still waters within brands, despite the carbonation process accounting for a significant amount of pre-bottling costs. Perhaps this illustrates how small these costs are in relation to the cost of packaging, transporting, marketing and retailing the stuff.

How big is the target market in a place such as Middle England? Assuming you drink two litres of mineral water a day (not including any for your tea and coffee) this works out at £142.35 to £394.20 per year for the leading non-premium brands, which is equivalent to 2.4% of average domestic water consumption by volume. But not 2.4% by cost, since while the average water and sewage bill would ease from £339 to £321, with 2.3 people per household the overall average household spending on water and sewerage would rise by £309–£889 – which is some difference.

The packaging costs more than the water. These costs, as with their environmental impact, have a habit of adding up.

Spunky water – Vidi, Vici, Veni?

Coca-Cola launched its Dasani bottled water brand in the USA in 1999 and in Canada the year after. The idea was to have a packaged water brand with a common name and image that could be bottled locally and developed for local market conditions worldwide. Soon the UK beckoned and Dasani's UK home page featured a smiling bikini-clad girl next to the slogan 'can't live without spunk' – a seminal failure to understand British slang. Then it emerged that the water came from a pipe provided by Thames Water, which was intriguing given that the entire aspirational side of drinking bottled water is that by spending more money you 'free' yourself from the purgatory of such plumbing. Dasani was marketed as using 'space age' treatment techniques. This turned out to be reverse osmosis through a semi-permeable membrane, something that has been commonplace in the sector since the 1970s. This is the triumph of the will concertinaed into the pressing concerns of reality.[17]

The water was retailed at 95 pence per litre against Thames Water's standard charge to customers of 63.27 pence per M^3 in 2003-04 – or a 1,502 times increase. In fairness, it may be noted that many media accounts stated it was being retailed for 95 pence a half litre. For the pedant, Thames Water estimates that its long-run marginal costs for water provision (at 2002 prices) is 50p per M^3 breaking down as 44p per M^3 for resources, 3p per M^3 for treatment, 2p per M^3 for bulk transport and 1p per M^3 for local distribution. At 95p per litre, it was clearly pitched at the mid-to-premium end of the market, despite its explicitly prosaic origins. As a bottled tap water, Dasani's obvious competitor ought to have been something along the line of Sainsbury's Basics range, whose two-litre bottles currently retail for 8p per litre. Indeed, it was pitched against the apparently super-premium Fiji Water. At 125p per litre, Fiji Water must have felt like a comparative bargain.[18]

The dream soon became a nightmare as tests revealed that Dasani water contained 10–22 parts per billion of bromate where it had not been materially present before.

The UK's Drinking Water Inspectorate's bromate limit is 10 parts per billion against the EU Drinking Water Directive's 25 parts per billion standard. Bromate is a by-product when naturally occurring bromide interacts with ozone during the treatment process. As the concentrations suggest, the greatest harm a habitual Dasani drinker faced was to their small change.

That this is not the way it came across is something of an own goal by the industry in general. This is because the essence of bottled water marketing has been to focus on people's fears about the risk involved in drinking safe, potable water, compared with various value-added products. We are often woefully ill-equipped to understand the nature of risk. Anybody who has gone through medical procedures such as the potential health of your as yet unborn child to any form of cancer prognosis will appreciate the visceral nature numbers take when we do not appreciate the nimble nature of risk. When people are enjoined to expect ever higher levels of perfection, reality can be a bit of a letdown.

This was not the first case of a premium brand being withdrawn on health grounds. Perrier found it had an image problem when bottles in North Carolina and Georgia in February 1990 were found to have 12.3–19.9 parts per billion of benzene against the Environmental Protection Agency's 5 parts per billion limit for public water supplies.[19] Perrier initially stated the problem arose from contaminated bottles being sent to North America. It transpired that the benzene was in fact naturally occurring in the gas used to carbonate the water and that the carbon filters used to remove it had not been replaced. 160 million bottles worldwide were eventually recalled. Later that month, the US Foods and Drugs Administration followed up Perrier's voluntary recall with a formal recall. The actual risk of getting cancer at this concentration is negligible and is in any case a fraction of what people living in urban areas inhale (without taking into account passive or active smoking), but fear does not respect statistics.[20] This is somewhat ironic, considering how many brands that are marketed via the debased principle of the fear of fear itself.

Bottled water is just as prone to health fears as tap water. Those who live in glass (or plastic) houses sow what they reap.

Reassuringly expensive?

Claridge's Hotel in London has its own water list, with prices worthy of the location. It is a wonderful place, Mayfair's Art Deco temple where the Bright Young Things swung through the Roaring Twenties, and I have happy memories of going to dinners and weddings many years ago in what now seems a more innocent age. Other hotels cater for those who liked to make a show of power or wealth; Claridge's was synonymous with well-connected discretion. Times change; in the pre-crash days of 2007 you could pay £21 for a 42 cl bottle of New Zealand's 420 Volcanic with a 'pleasant smooth sensation on the palate' as one would expect from a glass of water.[21] At £50,000 per M^3 this was getting about as good as it can get.

Elsenham jam was originally marketed as 'the most expensive jam in the world' and built a reputation for fine artisanal jams. The jam factory was located on the

Elsenham Estate on the Essex and Hertfordshire borders, where in 2000 Michael Johnstone was worried about a possible water leak as his water bills started to climb. Instead, he found a forgotten artesian well. The well had originally been opened in 1890 to supply the jam factory with water. The water was held in a 300-metre-deep chalk bed cut off from all other water sources. Tests by the Environment Agency found it was about as pure as spring water can get, and Mr Johnstone got out of the jam business and into water. As the web site[22] makes clear, Elsenham Artesian Spring Water is for 'the luxury premium end of the market' and distributed to 'only the very best' which means the 'very top hotels, restaurants, retailers and corporate boardrooms worldwide'. It is not usually available off the shelf, but in 2007 Claridge's sold it for £30 per 75 cl bottle, or £40,000 per M^3, while the Netherlands' Springwater Company offers it on the web for €5.25 per bottle. The fact that it comes from what is now an industrial estate in a less glamorous corner of England is immaterial. The water is as pure as its marketing and, since honesty is an admirable trait, I appreciate their approach.

The Springwater Company is a great source for the esoteric ends of bottled water. Amongst its other valuable offerings is Bling, which as its name suggests is pitched towards those who find comfort in labels and where everything must have a value rather than values. The standard 'Bling' comes in seven styles (packaging, the water remains the same) for €35 per 72 cl (before any mark-up), but the Dubai Series, launched with perspicacious insight in 2008, comes in a bottle encased with 10,000 Swarovski crystals at €2,500 a 72 cl bottle or £2,967,498 per M³ which truly defines a bubble economy.

Marketing designer water can be refreshingly direct in its aspirational tones. The web site Symrise has some entertaining accounts of what gives designer water its allure.[23] Tasmania's Cloud Juice has 9,750 drops of rain in every bottle, which is also known as 'Tears of God'. Although atheists will doubtless demand that it is re-labelled, there is no doubting the purity of this rainwater which as all rainwater ought to, consists of distilled water, no more, no less – and this rainwater is purer than most. According to the finewaters.com web site, Cloud Juice boasts a high 'virginality', although a reviewer from *The Observer*[24] found it to be 'chewy'. Even so, distilled water from a desalination plant rarely costs more than 0.1p per litre. Finé comes from the Shuzenji Hydrothermal Spring in Japan and is 'regarded as an especially pure water' even though the water 'has a constant temperature of 13° Celsius making it better able to absorb the mineral substances of the surrounding volcanic rock'. These waters are at the low end of Claridge's' list, the £9 and £15 bottles of Cloud Juice and Finé working out at just £12,000 and £20,000 per M³ respectively.

Since there are people who are happy to pay premium prices for all manner of premium-label drinks in such places, there is no moral objection to their knowing indulgence, assuming their money is the product of honest toil. However, as *Decanter* the wine magazine noted, a jug of (unadulterated) Thames Water won their mineral water blind tasting in 1999.

Every now and then, you get an insight into the underlying cost of premium water. For example, in

September 2011, I was offered a flexi-bag containing 24,000 litres of 'best glacier water quality!' from Iceland for delivery to a harbour of my choice in the Mediterranean for €9,600 or £350 per M^3.[25] Bottling and so on follows, but there is quite a jump from this to the price it could fetch in the right bar at the right time.

An elegant example of teasing at consumer gullibility came about when Boots the Chemists launched an 'Expert Sensitive Refreshing Facial Spritz' designed to protect the skin from dryness.[26] It costs £3 for 125 ml, a small wine glass, which works out at £24,000 per M^3. The sole ingredient is 'aqua' which is the cosmetic industry's euphemism for water, which, in this case, Boots claims has been 'specially formulated'. No matter what pre-treatment went on here, this water has been more 'specially formulated' in terms of its marketing rather than its physical chemistry. There are indeed other 'specially formulated' waters, such as heavy water or deuterium and tritium, which are used in the nuclear industry, but they are unlikely to be found on your average cosmetics counter.

If you wish to splash out on the premium brands and you can afford to, nobody should argue with you. But it would be good to consider what you are buying and why.

Value added burps?

OGO is water from a spring which is "complemented by a natural oxygenation process' meaning that Oxygenwater has 'significant additional oxygen molecules to boast an oxygen content of 200 mg/L which is 35X the oxygen content of regular water'. At £20 for twelve 330 ml bottles, this works out at 505,050 p per M^3. In fact, at room temperature (20° Celsius) at atmospheric pressure (sea level) there is 9.1 mg/L of oxygen.

The minor yet pressing problem is that we are mammals, not fish. We get our oxygen when breathing air which is 20.9% oxygen rather than from water. A single breath of air contains 100 ml of oxygen. When five oxygenated waters were analysed for a research letter to *The Journal of the American Medical Association* in November 2003, they had a calculated oxygen content of 2.5–22.6 ml of O$_2$ per 100 ml of water against 2.5 for tap water. This works out at 12.4–113.0 ml per half litre of water, assuming no oxygen is lost during opening, pouring and drinking. Given the time spent not breathing while glugging down barely a breath's worth of dissolved oxygen, it is unsurprising to note that in this trial, as with another reported in the *Journal of Exercise Physiology Online* in 2002, no beneficial effects were found when comparing exercisers who drank tap water to those who drank the oxygenated brands.[27] If you need more oxygen, you can save a lot of money simply by breathing in once more.

Some other approaches

Quite a bit of work has been done in seeking positive responses towards the challenges posed by bottled water. Often people buy water because they feel there are no alternatives in public places. Initiatives such as FindaFountain (aquatina.com) in the UK and WeTap (wetap.org) in the USA are aiming to provide people with locations of public water fountains. Some people are concerned about the possible danger of sharing a water fountain so the Give Me Tap (givemetap.co.uk) scheme which started off in Manchester and is being rolled out across Britain enables people who purchase a refillable bottle to present it at any of the participating cafes to get a free tap water refill. Part of the idea behind this is to take away the cringe factor that makes people feel obliged to buy bottled water when they are feeling thirsty.

Another approach has been developed by One Water (onedifference.org) where they sell locally sourced water in lightweight bottles. In the UK, the spring water comes from Radnor Hills, a spring in Powys, Wales. They also sell a filter jug for side-stepping bottled water at home. All the profits go to the One Foundation's water projects which to date have connected 1.5 million people to safe water while their lavatory products (from sales of loo paper and soap) are funding sanitation projects.

Carafes and cafés

I believe that one of the signs of a good restaurant is the cheerfully proffered carafe of chilled tap water on your table. If you wish to have some mineral water, that is your choice and that is fine, but the local water is happily provided. Of course, it is not free and, indeed, they may have put it past an in-house filter to give it a final polish, but you do not pay a specific supplement for electricity used or to defray the cost of staff uniforms. Tap water can be one of many commodities needed to spin the wheels of business and pleasure.

In 2005, Eau de Paris responded to a survey indicating

that 51% of Parisians drank bottled water by distributing 30,000 free carafes designed by Pierre Charpin. The idea was to show that when suitably chilled, their tap water tasted the same as the bottled variety. This stemmed from a legacy of French tap water either being seen as unhealthy or over-chlorinated. This campaign built upon a series of unrelated campaigns by Eaux d'Ile-de-France since 2002 to regain public acceptance of tap water. Between them, they parodied the health benefits, labelling and costs of the mineral water campaigns. While much lauded, there is little hard evidence that these campaigns have had any profound effect.[28]

Proprietors complain that mineral water is a seriously important cash cow. I would rather see a bill reflect the food and the service, and if there are any concerns about limiting a restaurant's income, then spend the difference on a better bottle of wine. Ultimately, all water and sanitation services benefit from transparency. An educated public deserves to be allowed to exercise a choice it is usually legally entitled to. With the data you need, informed choices can be made.

The better side of bottled water

Bottled water does indeed have a place on our tables, as something of a luxury item. Concerns about taste ought to push utilities to do better. The top end of the market is about generating revenues from people with serious money to spend and it has its own set of rules. More broadly, it is evident that European bottled water legislation is the way forward for the rest of the world.

Indeed, there is something rather noble about many of the smaller and localised super-premium brands. They bring money from prosperous urban areas to poorer rural areas, helping to sustain local economies. Because of their high prices, the value obtained for packaging and transporting this water is also higher. For example, Llanllyr Water in West Wales, which is abstracted and bottled on a certified organic farm run by the same family for 290 years, and their packaging has been developed with resource minimisation and recyclability in mind. The company told me that the farm set up its bottled spring water business in 1999 in order to diversify its income, to maintain its organic status and traditional farming practices and to boost local employment, with fifteen new jobs created by 2010.[29] Farming in Wales was under the cosh at the time, and fifteen jobs matter in an area such as Ciliau Aeron. This is joined-up thinking – revenues from the water help to sustain the farm's organic status which plays a part in assuring the integrity of the water source. If you're going to do something like this, then it is best to do that something well. If super-premium water can become a force for diverting discretionary spending towards the celebration of catchments and preserving their pristine environments, then that can be a real force for good.

Perhaps it is a case of investing in assets rather than adverts. Another important benefit of bottled water is that it may in time encourage people to treat water as something for which you need to pay a proper price. If bottled water has a value, so should tap water. Good catchment management, based on maintaining the integrity of the land in order to secure a high-quality water supply demonstrates that wholesome water carries an intrinsic value and one that has wider benefits. When water has a value, it can preserve values as well. This is where bottled and tap water can meet.

Conclusions – a question of values

Mens euill manners, liue in Brasse, their Vertues
We write in Water

William Shakespeare (1613)
The Famous History of the Life of King Henry the Eight

There is no compelling evidence to show that people are directly harmed by drinking bottled water but its aspirations of purity need to be treated with informed scepticism, unless, for example, it has met various EU standards. As a luxury, it is simply another beverage, but, as a necessity, bottled water ought to be seen as an indulgence.

Don't ban the bottled stuff – instead, educate people about all of its costs so that they can indeed make informed choices. Some clarity in labelling would also be welcome, since, as we have seen with attitudes towards cigarette packaging and advertisements over the past 60 years, perceptions can change. Can other people choose what water they drink and should people be allowed to choose what water they drink? The same applies for their sanitation services. This leads to the question: why is the public disconnected from these processes?

Spring and mineral waters can encourage utilities to polish up their act, along with the use of point-of-use filters and filter taps at the kitchen sink. There is a case

303

for apartment blocks and housing developments investing in a water unit to smarten up the drinking water. All these things could help make bottled water (and carbonated drinks) become an exception rather than a desired norm.

As with mobile telecoms, there is the matter of choice, but it needs to be an informed choice. People can choose to pay what hotels bars and restaurants charge for their designer waters or they can spend their money elsewhere. Yet there is something terribly sad about deliberately paying over the odds just to be seen to be doing so.

If you buy bottled water, you can make your money go further when bottled water companies support water charities. Llanllyr Water for example, gives a proportion of its profits to Just A Drop (justadrop.org) who have provided water to 1.5 million people in 29 countries. All of One Water's (onewater.org.uk) profits go to the One Water Foundation (onedifference.org), connecting 1.5 million people to safe water, while AquAid, a water cooler company, supports WaterAid (wateraid.org) and Pump Aid (pumpaid.org). As well as supporting WaterAid, Belu (belu.org.uk) is involved in We Want Tap (wewanttap.com) a web-based campaign against bottled water. I am an Ambassador for Pump Aid; their locally built and maintained Elephant Pumps serve 1.1 million people in Africa. WaterAid supplies 15.9 million people with clean water and 11.0 million with safe sanitation since it was founded by UK water utilities in 1981, companies that have raised £115 million for WaterAid in the past 30 years. Other UK based charities active in the field include Cafod (cafod.org.uk), Christian Aid (christianaid.org.uk) and Frank Water, who have connected 400,000 people in India (frankwater.com). For tens of millions of people, these charities do make a difference and deserve support, sharing a commitment to delivering rather than prevaricating.

Likewise, we are told we need to constantly hydrate. This is also called having the occasional glass of water or cup of tea or something similar at sensible intervals through the day. There is no desperate need to clutch to a bottle of water each time we cross the road or walk into a meeting. Marketers are able to make infantile assumptions about people's understanding about basic science because, as with the 'dihydrogen monoxide' spoof we saw in chapter one and attacks on scientists over climate change in chapter three, ignorance about, or indeed contempt, towards science currently is a career-enhancing capability for many. Oxygenated and super-oxygenated water might, in theory, induce a value-added burp. In reality, the only change will be in one's spending, as money goes towards a product that has no material difference to tap water.

The forecast of $100–180 billion being spent worldwide on bottled water by 2014 has a rounded feel to it. Imagine that money being unleashed on water and sewerage service extension worldwide. Money not spent on bottled water could be transformed into new funding for water and sewerage infrastructure either at home or abroad. Of course, that would not happen for many reasons, but it highlights the choices we make and their implications.

[1] Hill R (2011) Real Water – with added electrons, *Guardian Notes & Theories*, 21st April 2011, Guardian.co.uk/science/blog/2011/apr/21/real-water-added-electrons

[2] Datamonitor, Bottled Water: Global Industry Guide 2010, http://www.datamonitor.com/store/Product/bottled_water_global_industry_guide_2010?productid=4FC1557E-937D-4783-8325-4B95025BAB19
Euromonitor, from *Financial Times*, 8th December 2010, Emerging markets tap into trend for bottled water http://www.foodbev.com/report/global-bottled-water-driven-by-developing-markets
Beverage Marketing Corporation (2009) *The Global Bottled Water Market*, BMC, New York, USA

[3] http://www.britishbottledwater.org/vitalstats2.html

[4] McCarthy D (2009) BBC accused of wasting £406,000 of public money a year on bottled water, *The Guardian*, 11th August 2009

[5] OECD Observer 278, March 2010

[6] Williams E (2001) Just add water, *Times Educational Supplement*, 1st June 2001, Tes.co.uk/article.aspx?storycode=347651, accessed 25th April 2011
Williams E (2004) Water in schools, *Times Educational Supplement*, 21st April 2004, Tes.co.uk/article.aspx?storycode=394947, accessed 25th April 2011

[7] http://www.washingtontimes.com/news/2008/aug/19/bottled-or-tap/
Saleh M A et al (2008) Chemical, microbial and physical evaluation of commercial bottled waters in greater Houston area of Texas, *Journal of Environmental Science and Health*, A Tox Hazard Subst Environ Eng. 2008 Mar; 43(4):335–47

[8] Misund A, Frengstad B, Siewers U & Reimann C (1999) Variation of 66 elements in European bottled mineral waters, *Science of the Total Environment*, 1999 Dec 15; 243–244:21–41
American Society for Microbiology (2010, May 26) High level of bacteria found in bottled water in Canada, ScienceDaily, Retrieved March 15, 2011, from ttp://www.sciencedaily.com/releases/2010/05/100525140954.htm
Nouvenne A et al (2010) Effects of a low-salt diet on idiopathic hypercalciuria in calcium-oxalate stone formers: a 3-mo randomized controlled trial, Am J Clin Nutr March 2010 vol. 91 no. 3 565–570

[9] Gleick P H (2004) *The world's water 2004-05: the biennial report on freshwater resources,* Island Press, CA, USA

[10] Walker A. (1992) Drinking water-doubts about quality, *BMJ*, Jan 18; 304 (6820):175–178

[11] Council Directive 80/778/EEC of 15 July 1980 relating to the quality of water intended for human consumption and as amended

in Directive 98/83/EC. For bottled waters, Directives
80/777/EEC, amended 96/70/EC and then recast as 2009/54/EC
The Directive 2003/40/EC and Commission Regulation 115/2010
also apply

In the UK, these are defined under 'The Natural Mineral Water,
Spring Water and Bottled Drinking Water (England) Regulations
2007 (as amended)' and administered by the Food Standards
Agency

[12] Gleick P H (2004) *The world's water 2004–05: the biennial report on freshwater resources*, Island Press, CA, USA

[13] http://www.foodbev.com/report/bottled-water-container-recycling-rate-doubles

[14] Gleick P (2010) *Bottled And Sold: The Story Behind Our Obsession With Bottled Water*, Island Press, Washington DC, USA

[15] www.sainsbury.com, accessed 7th March 2011

[16] Prices from mysupermarket.co.uk comparison site, accessed 7th March 2011

[17] http://www.theregister.co.uk/2004/03/11/introducing_dasani_the_water/, accessed 27th September 2010

[18] Ofwat (2004) Water *and sewerage service unit costs and relative efficiency, 2003–2004 report*, Ofwat, Birmingham, UK
Ofwat (2003) *Tariff structure and charges, 2003-04 report*, Ofwat, Birmingham, UK
http://www.timesonline.co.uk/tol/news/uk/article1049413.ece, accessed 27th September 2010
Binnie C, Kimber M & Smethurst G (2002) *Basic water treatment (3rd edition)*, Thomas Telford Ltd., London, UK

[19] Borman S (1990) Benzene in Perrier found by North Carolina lab. Chem. Eng. News 1990 68 (8) 5–6

[20] Cocheo V et al (2001) Urban Benzene pollution and population exposure, WHO Collaborating Centre for Air Quality Management & Air Pollution Control, Berlin.

[21] http://www.decanter.com/news/wine-news/486276/claridge-s-to-sell-water-at-50-per-litre, accessed 28th September 2010

[22] www.elsenhamwater.com

[23] www.always-inspiring-more.com/Some-luxury-waters.134.0.html, accessed 14th October 2010

[24] Davis J (2003) Would madam care to taste the Cloud Juice?, *The Observer*, 2nd December 2003

[25] Email from Ulf Leonard, Leonard Ventures, 16th September 2011

[26] www.timesonline.co.uk/tol/life_and_style/health/article 2599566.ece, accessed 13th October 2010 'It is gentle and refreshing. It is £32.92 a litre at Boots. It's water', *The Times*, 6th October 2007

[27] Hampson N B, Pollock N W, & Piantadosi C A (2003) Oxygenated water and athletic performance. *Journal of the American Medical Association*, 290, 2408–2409

Willmert N, Porcari J P, Foster C, Doberstein S, & Brice G (2002) The effects of oxygenated water on exercise physiology during incremental exercise and recovery, *Journal of Exercise Physiology Online*, 5, 16–21

As described by bodykind: www.bodykind.com/product/26_40-OGO-Oxygenated-Water-12-x-330ml.aspx, accessed 14th October 2010

www.ogolife.com, accessed 14th October 2010

[28] http://news.bbc.co.uk/1/hi/world/europe/4373205.stm, accessed 13th October 2010

Parisians tempted with tap water, BBC News, 22nd March 2005 http://www.sircome.fr/?Promoting-Tap-Water, accessed 14th October 2010

[29] www.llanllyrwater.com, accessed 14th October 2010

Chapter 8

Fear, Loathing and Finance

People expect to pay for goods and services, yet water is somehow different. In England and Wales, if you walk out of a shop with an armful of shopping without paying, you expect to be prosecuted if caught, and if you don't pay for your mobile phone service, you don't expect it to work. But if you don't pay your water bill, thanks to legislation passed in 1999, you will continue to be served, subsidised by customers honest (or naive) enough to carry on paying. Water is unusual across the world. A survey by the World Bank a decade ago found that 39% of water tariffs were inadequate for any cost recovery, 30% covered some of the operating costs and 30% covered some of the capital costs.[1] This was closely linked to wealth, with the proportion paying at least part of their operating costs ranging from 12% in low-income countries to 92% in high-income countries and 69% for electricity in low-income countries.

Tariffs: Full Cost Recovery and Sustainable Cost Recovery

When it comes to paying for water and sewerage services, there are three main elements: the cost of operating the system ('operations & maintenance' or OPEX), the cost of new or rehabilitated equipment and systems (capital expenditure or CAPEX) and, if applicable, the cost of servicing debts held by the utility. This is the 'blue water' side of water management that is usually associated with Full Cost Recovery, dealing with water that the utility uses. Beyond this, there is also a debate about who ought to be responsible for managing and paying for 'green water', the surface water that sustains terrestrial life in general but is not used by the utility.

Full Cost Recovery is something of a misnomer, since apart from Denmark, which also has the world's highest tariffs, no utility actually recovers all of its costs.[2] In reality, this means a utility that covers all of its operating costs, a part of its capital costs, and is able to pay for the rest of these through raising and servicing debt finance. As utilities are rarely directly responsible for the 'green water' side of environmental management, an element of water management spending is usually carried out by the state. For example, in England and Wales, the Environment Agency is responsible for 'green water' while monitoring the performance of the privately owned utilities to ensure that their 'blue water' activities do not impair these.

In Denmark people pay for what they get. Even the rainwater that flows from Copenhagen's drains is subjected to tertiary treatment before being put back into the water cycle. High costs are seen as part of a programme for comprehensive sustainable Full Cost Recovery. This outlook lay behind the European Union's desire that from 2010, as part of the Water Framework Directive, Full Cost Recovery would be the norm in all 27 Member States. Given that many member states have yet to consider how much this directive will cost, this remains a work in progress.[3]

Jim Winpenny's *Camdessus Report* introduced the concept of Sustainable Cost Recovery.[4] This explicitly recognises the difficulties in financing both the blue and green water management using Full Cost Recovery, especially when finance is expensive or in developing economies, where affordability is going to be a challenge. Sustainable Cost Recovery aims for funding through the 'Three Ts' or tariffs, taxes (government support) and transfers (international and local aid), with predictable flows of public support and affordability built into the bargain. Further funding through debt has to be repayable. In many developed countries, the taxes come as water abstraction and pollution charges and sometimes there are specific charges for household water consuming devices as levied in Singapore.[5] In November 2011, the UK Government announced that from 2013, customers of South West Water would get a £50 rebate from the Government to offset their water bills. Currently the average customer bill in their area is £519 compared with an average bill of £319 for Thames Water.

Surveying funding sources by 67 countries in 2009, I found a very wide range of funding mixes, from 60% of work being funded by tariffs (Singapore and northern Europe) to 60% being funded by debt (Japan) to 100%

coming from the state (Qatar) and 40% coming from European Union grants (Estonia and Latvia).[6] Personal wealth also matters. The World Bank considers affordability to be an issue when bills for water and electricity exceed 5% of household income when covering operating costs, which would affect 20% of households in Latin America and 70% in India and Africa.[7] Surveys look at utility bills as a percentage of household income and viewed this way most water bills are relatively small, especially when compared to electricity and telecoms bills, ranging from 0.2–0.3% in North America, 0.2–1.0% in Western Europe and 0.3–1.0% in developed Asia.[8] The OECD's 2007–08 survey is a useful snapshot of tariffs and their affordability in member states.[9]

Domestic tariffs and water bills as a % of household income for average and poor households

	Tariff $ per M³	% of average Household income	% of poorest 10% Household income
Mexico	0.49	0.3%	4.2%
Japan	1.85	0.3%	1.4%
Poland	2.12	1.2%	7.9%
Sweden	3.59	0.5%	1.8%
France	3.74	0.7%	2.2%
Denmark	6.70	0.9%	2.5%

It is interesting to see that tariffs in the more unequal societies (Mexico and Poland) have much more impact amongst the poor than do more equal societies such as Denmark and France, despite their tariffs being appreciably higher. For the poorest 10% of households, it is clear that affordability matters. Here, the share of the household bill should not exceed 3% in developed economies and 5% in developing companies and the second 'T' (taxes) when used as targeted subsidies ought to be used to minimise this burden.[10] People need water and will pay for it, with families in Zambia willing to pay 3–10% of their household income for a safe supply, which is less than they currently pay to informal vendors.[11] Indeed, there is plenty of evidence that targeted financing works when pro-poor pricing is combined with services being extended to them. In Casablanca and Singapore, LYDEC and the PUB use social and rising block tariffs, where water used for non-essential purposes such as baths and car washing costs more than water used for essential purposes and subsidises the latter. Other successful approaches include tokens for paying water bills (Macao Water, Macao), direct subsidies (Aguas Andinas, Santiago) and low-cost household connections (Manila Water, Manila). This is something of a balancing act, since if you 'lose the first block' by providing a quantity of free or too cheap water each month as in South Africa, you can lose too much revenue when there is only so much available for cross subsidies.

How much needs to be spent?

In 2011 I was invited by the OECD to look at how much needs to be spent on water and wastewater infrastructure up to 2050. The Organisation for Economic Co-operation and Development was founded in 1961 as a policy forum building upon the US funded Marshall Plan which from 1947 had been instrumental in rebuilding Western

Europe's war-torn infrastructure. One of the Plan's achievements was to develop a comprehensive urban sewerage network for the Netherlands and the OECD's interest in water continues.

Previous spending surveys show that estimates can be an imprecise art

People like numbers, especially when they tell you how much *is* and *ought* to be spent on something. In England and Wales, annual data produced by Ofwat, the sector's regulator, gives a precise figure for how much was spent on various aspects of developing and managing water and wastewater assets. Thanks to Ofwat's five-year regulatory programmes, you can make some pretty useful forecasts. In most countries, this is simply not the case, especially when it comes to countries that have no idea about how much is being spent – let alone what needs to be spent. Between 2000 and 2004, at least eight sets of forecasts were made about the cost of the water and sanitation MDGs, estimating the cost at between $6.7 billion pa (a basic service) to $72 billion pa for a fuller service. To halve the proportion of people without a safe water supply by 2015, an estimated $2 billion to $23 billion per year would be required, depending on the approach taken, while for basic sanitation, $2 billion to $17 billion would be needed per year. The World Water Vision anticipated $1,800 million would be needed on the basis of $500 per person newly connected, but excluding sewage treatment.[12] Other forecasts came in at $75–100 billion per annum.[13]

314

A survey of these surveys was carried out by the environmental consultants ERM for the UK government's Department for International Development (DfID) in 2003[14] and noted a range of estimates from \$9–180 billion per annum as being needed to meet all these targets, describing one set of estimates as being 'scarcely more than "back of the envelope"' in their nature, as did a review by the United Nations which noted that 'clearly, all the above cost estimates are "back of the envelope" calculations'[15] – which is a pity, considering the scale and urgency of the task. Again, there are issues regarding definitions. Some surveys include operations and maintenance costs and overheads (administrative expenses) while others only look at actual construction costs. Meanwhile estimates by NGOs such as the UK's WaterAid are often lower than official estimates because they focus on minimising overheads and installation costs through using local labour.

Guy Hutton and Jamie Bartram made a useful analysis of spending needs for the World Health Organization in 2008.[16] Assuming individual household connection costs of \$148–232 per person (in 2005 values) for water and \$193–258 for sewerage and at the 'community' level \$50–72 for water and \$93–134 for sanitation, they estimated the annual cost of new urban water connections at \$15 billion and sanitation connections at \$51 billion over ten years, along with \$235 billion needed to maintain and rehabilitate extant urban water assets and \$158 billion for sewerage systems.

Globally, estimates of \$400–1,000 billion being spent each year on all aspects of water worldwide have been managed in a most 'eco-friendly' manner over the past decade, by being uncritically recycled. These guesses have not changed very much since perhaps 2000 and neither has actual spending in developing economies. One study pointed out that capital spending in developing economies needed to jump from $30 billion pa in 1995 to $82–92 billion pa in 2000–25, with a major emphasis on sewerage and sewage treatment. It is pretty clear that no such jump has taken place, especially for sanitation, the sector's 'poor cousin'.[17] For real progress, real data is needed about what is actually being spent and where.

Recent studies have highlighted spending needs, but do not relate to what is being spent today. Lisa Saroli, at the Ontario Ministry of Economic Development and Trade, believes C$22,000 billion (US$21,800 billion) will be spent between 2011 and 2030[18] on 'water scarcity solutions'. In 2007 management consultants Booz Allen Hamilton unveiled a set of infrastructure spending needs for 2005–2030 – including $22,600 billion for water, exactly 2% of which ($226 billion each) was earmarked for the Middle East and Africa against $4,972 billion for Latin America.[19] While the African figure seems on the low side, the probability of $7,500 being spent per capita on water and wastewater in Latin America during this time looks rather ambitious. As with forecasts by Deutsche Bank of water investment needs of €400–500 billion pa, headline numbers such as these get a lot of publicity but have limited value when detailed breakdowns are not available.[20]

Global Water Intelligence's *Global Water Markets 2011* may be a useful development here.[21] Christopher Gasson as GWI's publisher has sought to get data out into the open so that at least it can be queried and improved upon.

This work was developed by a team of correspondents and sector specialists (myself included) combining 50 country studies and a top down look at 163 countries in all. They estimate that in 2010 $508 billion was spent on water and wastewater goods and services worldwide, including $28 billion by industrial clients and $396 billion by utilities.

Their municipal capital spending estimate of $173 billion ($90 billion on water and $83 billion on wastewater) is seen as little more than allowing assets to deteriorate in an orderly manner as possible. To keep at a standstill, they believe that $211 billion pa is needed, $83 billion to maintain existing assets, $38 billion to meet population growth, $60 billion for complying with regulations and a further $30 billion to secure supplies. When it comes catching up with decades of underspending, they suggest a further $260 billion is needed every year, including $83 billion rehabilitating worn out networks, $67 billion on extending infrastructure, $180 billion on meeting new regulations worldwide and a further $30 billion on security of supply. Perhaps the real value here lies in the implied change in spending levels to move from [1] business as usual in terms of managed decline to [2] holding our own and [3] to securing a universal and high-quality service.

A CEWWT little model

Funding fascinates me. I developed a model, dubbed CEWWT (Comprehensive Evaluation of Water & Wastewater Treatment) as part of writing two bone-dry books on water finance in 2006 and 2009.[22] CEWWT sought to estimate how much money was needed in developing new and rehabilitating existing assets over a twenty-year period for 59 (2006) or 67 (2009) countries based on realistic expectations about what would be spent. With a 40-year timescale and an emphasis on what could and should be done in 156 countries during that time, 2CEWWT, its successor could afford to be more ambitious.

In developing 2CEWWT, my guiding principle was that by 2050, everybody ought to have continual access to safe water and sanitation and that wherever circumstances allow, that access is enhanced to optimise the quality of people's lives and to ensure basic steps are taken towards maintaining the integrity of the water cycle. A 40-year projection ought to be optimistic in a way the millennial initiatives could not be. A 'full' coverage scenario was developed where all people living in urban areas have access to piped water and household sewerage. In the 'basic' scenario, half of those living in slum areas get local access to a continual supply of potable water and fully managed sanitation blocks, while the other half have full household coverage. In the 'basic' scenario, half of all sewage in developing countries is fully treated and half the population has storm sewerage. I hope that the 'full' rather than the 'basic' scenario will prove to be the more accurate forecast.

Cost elements needed for universal coverage were developed for bulk water distribution and treatment, household water delivery, household sewerage and sanitation, mains sewerage and sewage treatment and recovery, along with water metering, systems monitoring and where appropriate desalination and advanced water and sewage treatment. High-cost countries such as Austria and Germany have been treated as exceptions rather than the norm in order to emphasise affordability.

Even complying with a single law is a complex process. The European Union's Urban Wastewater treatment Directive calls for all sewage from urban areas to be properly treated by 2005. While Germany, Austria and Denmark spent €500–1,200 per person to meet the directive, others like the Netherlands, UK, France and Spain spent just €150–270. The fact that Spain and France were still struggling to comply in 2011 shows that it's not just spending but spending well that matters.[23] At an estimated cost of €155 billion by 2005, it has not been a cheap exercise, but Europe now has a modern sewage treatment infrastructure.[24]

2CEWWT's estimates are not comprehensive as they do not take into account areas such as Climate Change (studies into costs are emerging), the EU Water Framework Directive (a work in progress), new drinking water / environmental contaminants (as we saw in chapter one, it is a new area) and sustainable storm sewerage (an old area due for a revival).

Rehabilitating water and sewage treatment works usually takes place every 20–30 years although some sub-systems and components need more frequent replacement or overhaul.[25] Networks and treatment facilities do need to be replaced when they are worn out or outdated, but there is no fixed operating life here as every network has its own pressures whether the nature of the water or the lie of the land. Water and wastewater distribution systems are designed to last for 40–80 years (but this can be decades longer). Networks do deteriorate with time and, in the UK, Thames Water has to deal with 35–50% distribution losses from pipes laid 120–150 years ago in north London.

Country information about the best and best value ways of installing new water and sanitation services was looked at in some detail and I developed estimates for the cost of installing things such as water pipes and mains in different economies.[26] While the cost of advanced treatment systems may be broadly the same the world over, costs of labour and materials clearly vary – as do the costs of complying with various national and regional drinking water and environmental standards. Advanced water and wastewater treatment are primarily concerned with responding to regulatory standards, along with global standards and expectations for utilities and companies operating in a multinational context. Systems metering and monitoring refers to developing appropriate reporting systems for the optimal management of water and wastewater utilities. Again, this is biased towards more developed economies and cities for the next few decades.

Being something of an idealist, I have assumed good practice with regards to procurement and corruption. This means that the costs here reflect the cost of goods competitively ordered and installed and that the money used here goes on the projects, rather than being directed to other parties. As Transparency International noted in their 2008 survey of corruption in the water sector,[27] this is often not the case and while data is poor, these factors can distort project costs by 10–40%.

These figures are a useful first attempt at quantifying capital spending needs. The model was developed with flexibility in mind so that all new information can be incorporated in order to improve its accuracy.

2CEWWT: Summary forecasts of capital spending needs, 2010–50 – Part 1

Water & WW infrastructure ($ million – 2010 values)	Water distribution	Water treatment	Sewerage	Wastewater treatment
Regional total				
Sub-Saharan Africa	181,721	37,767	369,238	357,096
Asia & The Pacific	320,477	385,037	1,008,879	689,974
Eastern Europe & C A	109,238	41,841	204,091	131,837
Latin America	123,341	59,612	305,352	179,418
Middle East & N Africa	101,665	156,824	256,913	145,948
North America	238,603	145,182	484,641	256,721
South Asia	274,311	96,228	611,252	422,138
Western Europe	120,394	106,556	289,904	183,129
Global total	1,469,750	1,029,045	3,530,270	2,366,261
World Bank status				
High income	420,562	337,970	956,864	561,813
Upper mid income	168,671	88,162	398,181	249,631
Lower mid income	597,422	540,048	1,598,297	1,050,897
Low income	283,096	61,953	576,928	503,919

2CEWWT: Summary forecasts of capital spending needs, 2010–50 – Part 2

Monitoring & totals ($ million – 2010 values)	Metering & monitoring	Water & WW – basic	Water & WW – full
Regional total			
Sub-Saharan Africa	21,912	587,649	967,734
Asia & The Pacific	91,079	2,022,979	2,495,446
Eastern Europe & C A	25,839	496,904	512,844
Latin America	21,827	561,419	689,550
Middle East & N Africa	16,100	562,116	677,451
North America	30,308	1,155,453	1,155,453
South Asia	28,467	1,006,132	1,432,396
Western Europe	41,048	741,030	741,030
Global total	276,578	7,133,682	8,671,904
World Bank status			
High income	93,672	2,359,956	2,370,573
Upper mid income	39,753	840,539	944,437
Lower mid income	111,040	3,006,338	3,897,704
Low income	32,114	925,669	1,458,010

Mind the gap

Estimating the size of a market and forecasting future spending needs calls for context – how much money is actually available to finance this work? In heavily regulated countries such as England and Wales this data is easily available and pretty accurate, but many countries have a poor grasp of both utility income and expenditure. Two attempts have been made to take a look at this on a global scale, plus many more national and regional studies.

A survey by Global Water Intelligence looked at water and wastewater capital and operating spending in 163 countries in 2007 and compared them with their estimates of their utility revenues.[28] Utility revenues were estimated at $234.1 billion and operating costs at $183.3 billion. That points to an operating surplus of $50.8 billion. Capital spending was estimated at $141.6 billion or a shortfall of $90.8 billion. This does not take into account the costs of servicing debts and is at a time when global water spending is neither meeting the Millennium Development Goal targets nor making any real progress towards sustainable water management worldwide. It is also at a time when many networks and treatment plants are deteriorating due to a lack of maintenance spending, let alone spending on replacement and rehabilitation. When I applied tariffs that I thought could be achieved in 67 countries and set them against some pretty basic spending targets for 2010–29, I ended up with a global funding shortfall of $1,049–2,297 billion a gap of $52–115 billion per annum.[29]

Time does not stand still when maintenance spending is frozen. In the USA, 88% of sewers were seen as being in good or very good condition in 1980 and 7% as very poor or worse.[30] By 2000, 60% remained in good condition and 11% needed immediate attention. Without new spending to address the current infrastructure, 44% are expected to be in good condition in 2020 and 32% in serious need of replacement. Little has taken place since 2000 to change this view. As a result, the Clean Water[31] (the US term for sewerage) spending needs identified rose from $221 billion in 2000 to $298 billion in 2008 (in 2008 dollars) including a 35% jump in sewer replacement and rehabilitation, and Drinking Water[32] needs identified rose from $200 billion in 1995 to $335 billion in 2007 (in 2007 dollars), although the earlier survey under-reported various problem areas.

Bridging the funding divide

How can we finance the spending gap? The current financial crisis is making matters harder, as much new money came from issuing bonds and long-term loans. There is some hope here as water bonds are a safe investment, since municipalities usually make sure they get repaid. In the USA, just 0.03% of municipal water and sewage bonds went into default in 1980 to 2000 against 0.5% of AAA (the top rated) corporate debt during that time.[33]

Sophie Tremolet, a water economics consultant, has been looking at financial mechanisms and financial innovation for some time.[34] Her study for the OECD provides a useful overview of the many mechanisms that have been developed to try to finance water work equitably and sustainably.[35] The study also highlights what a challenging subject this continues to be. Here are some examples.

In the USA, municipalities can also get funding through revolving credit facilities, where funds used to repay Federal loans are used to provide new finance. Since 1987, most Federal government spending comes via its Clean Water State Revolving Loan Funds (CWSRF) programme and in 1996, a similar programme, the Drinking Water State Revolving Loan Funds (DWSRF) was adopted. These State Revolving Loan Funds encourage both individual states to invest in their own infrastructure by providing matching funds and, where appropriate, with further funds raised through issuing tax-exempt bonds to retail investors and fund institutional investors. Because the debt is guaranteed and it is in a low-risk business, funding costs are a third lower than normal. The Clean Water fund has mobilised $70 billion in finance in twenty years and the Drinking Water fund a further $14 billion in its first decade.[36]

Korea funds municipal and national water projects through Kowaco (Korea Water Resources) that are repaid through tariffs, and Kowaco enjoys a higher quality debt rating than the Korean Government.[37] Other national water debt funds can be found in Japan and the Netherlands, amongst others. Companies and Governments are active in the debt markets, with $62 billion of water related bonds being issued in 2000–09 and $156 billion of syndicated loans issued for water and waste management activities worldwide.[38] The development banks, which range from the World Bank (funds to developing countries) to the European Investment Bank (funds to European companies and countries), provide loans in the region of $5.5 billion each year.

Below the national level, unless a state or municipality can get itself rated by one of the debt agencies (chiefly Standard + Poor, Moody and Fitch), raising funds is an expensive task, as the riskier a project is seen to be, the higher the interest rate will be charged. This is a serious challenge when it comes to raising new funds at the sub-sovereign (below the national government) level in developing countries. Over the past decade, a number of initiatives have emerged to address this. In structured funds a guarantee facility is raised by an agency or development bank against the project risk and this has been used in Tamil Nadu, India.[39] Pooled finance allows a number of small projects to lower their individual finance costs and obtain longer-term repayment periods. In the Philippines, the Land Bank of the Philippines' Water District Development Project pooled thirteen projects with loans ranging from $0.4 million to $7.6 million with project risk being spread through the group, while the Local Government Unit Guarantee Corporation provides a guarantee that loans will be repaid for a guarantee fee costing 0.5% to 1.25% of the loan.[40] Colombia's FINDETER[41] is a collective government-backed vehicle which provides low-cost, long-term debt to municipalities and supported 300 water and sewerage projects alone in 2005; this was followed by Grupo Financiero de Infraestructura Ltda, raising $62.5 million in internationally rated bonds in 2009 for smaller-sized municipalities.[42] These are all small and local, but they can and ought to be adopted elsewhere.

Guarantees facilities can do great things when it comes to raising funds for projects where the benefits may not come though in purely monetary terms. Between 2006 and 2010 the International Finance Facility for Immunisation raised $3.08 billion in ten bond issuances.[43] $1.73 billion had already been disbursed by the end of 2010 and six donor countries have guaranteed providing $5.98 billion in funding to repay the bonds over twenty years. The fact that these payments are guaranteed by these governments means that the bonds are AAA rated, minimising the cost of finance. The funds are provided to the GAVI Alliance that aims to immunise 500 million children through the $4 billion to be raised.

Official Development Assistance covers international aid from developed countries to developing countries. It is often assumed that developing countries need to depend on international aid to develop their infrastructure. This is as patronising as it is incorrect, not least since water is a low priority for donor countries. During the 1990s, these funding sources actually fell before making a recovery in recent years.[44] In 2002–05, funds disbursed for large scale and basic water and sanitation projects ran at $1.3 billion per annum, rising to $2.5 billion per annum in 2006–09.[45] While all aid is welcome, these levels of support will make little impact in extending water and sanitation services. Aid is too often used as a blunt instrument, sometimes doing more harm than good. New approaches such as output-based aid (OBA) target funding to where it can make the greatest impact and focus strongly on appreciating local circumstances.[46] Aid also works best where it leverages other investment, often by concentrating on making hardware such as a septic tank affordable.[47] Building a sanitation block and walking away is a waste of everybody's money. Hygiene promotion is essential for their beneficial use.

The World Bank has been involved with water and sanitation projects under various models since the late 1960s and has consistently found that capital-intensive projects funded in a hard currency and paid for by sort currency tariffs can be an accident waiting to happen.[48] Problems repaying loans after currency devaluations have undermined projects ranging from Buenos Aires (Aguas Argentinas) to Manila (Maynilad Water). Between 1990 and 2010, the World Bank supported 731 projects involving private sector participation. By 2010, 63 of these had been cancelled or were in distress and more tellingly these covered 33% of the $62.5 billion of the project financing. In contrast, 3–8% of projects by value in the World Bank supported private sector projects for electricity, telecoms or transport were in distress or cancelled by 2010.[49] The fact that Full Cost Recovery for these other sectors is the norm worldwide must have something to do with this disparity. There is some scope for encouragement, since the World Bank's overall water portfolio has been improving. In 1995, 49% by value of all World Bank water and sanitation loans[50] were at risk or had been cancelled, compared with 28% for infrastructure projects in general. By 2005, 9% of water and sanitation projects were in a similar state, compared with 10% of infrastructure projects overall.

All of this is encouraging, but it can only address a fragment of the current and future funding needs. Because of our current debt crisis, it is unlikely that many governments will be able to afford increased subsidies for some years, so water and sewerage needs to become more finically self-sufficient and more attractive to financiers able and willing to provide the type of funding needed.

The moral case for paying for water

So, how does the financing of water and sanitation services affect human welfare, happiness and sustainability? And what happens when we are confronted with the need to recognise the value embedded in our water resources and the costs of mobilising them for beneficial use?

I believe there is a moral case for using Full or Sustainable Cost Recovery as part of the process of tackling the manifest shortcomings in every level of water management today. If water is 'free', who will provide it? What incentive does a municipality have to provide such a service, especially to those whose political and economic influence is seen as being marginal? If the funding flows are not there and the foreign donors do not step in (as usual), this service simply will not be provided and as we have seen, it usually is not provided. The figures for non-connection, especially in slums, highlighted in chapter one, demonstrate the sheer level of this neglect.

One of the reasons that no funding is available is because politicians and policymakers labour under the assumption that people cannot afford to pay for these services. This is an extraordinary and dangerous nonsense on many levels. People need water irrespective of their circumstances and have to obtain water in order to survive whether the municipality chooses to supply it or not.

Where do those people the municipalities overlook get their water from? Since they are not meant to be able to afford to pay, presumably it is free. Yet in the same way that they pay for food and clothing, they most certainly do pay for water and sanitation. Without formal provision of water, especially at the household level, you either have to rely on the suspect offerings of unregulated entrepreneurs who step in where the municipalities do not care or take a chance from contaminated wells and the such-like. Either option involves massive costs. Water entrepreneurs run small-scale operations that are highly labour intensive, with all the costs, especially their onerous cost of capital (informal finance involves three-figure interest rates while utilities agonise over fractions of a per cent) being passed on to the end users, a case of Full Cost Recovery with a mighty premium added. The urban poor pay vendors up to 240 times more for their water than the wealthy pay in the same cities for their water. Where is the moral case for letting this happen? In effect, by not being served, the urban poor are subsidising the urban rich, which goes some way to explaining why there is such an eloquent and well-funded lobby to retain the status quo worldwide. Of course, when paying so much more, you use less, but you lose the several benefits of enjoying a proper water service. And those who have to rely on unprotected sources? The figures on mortality and time lost being too ill to work amongst the un-served tell their story. Any area's 'natural' water resources have their limits and anybody who expects a household 'gift from God' of potable water is guilty of both a selective reading of the Scriptures and needs to appreciate what exactly is going on in the real world today.

The responsibility for ensuring that water and sewerage services are properly and sustainably funded rests with a nation's government. The role of government at the national and local level ought to be about serving its citizens. Ensuring full and safe access to water and sanitation ought to be one of the prime reasons we have these administrations since without them, no government can be regarded as supporting a just and reasonable civic society.

Ensuring that these services are delivered efficiently and equitably is the duty of the utility and of government in overseeing the utility. The lower the cost of providing these services, the lower the tariffs need to be. That is why the prime duty of a utility is to deliver these services, rather than to maintain excessive staffing levels and play a role in local political patronage. For utilities run by the private sector, they have a moral imperative to provide the best service for the best price as this allows contracts to run for a long period of time within a framework of trust between the government, the company and its regulator.

There are arguments that informal water vendors are a good thing as they create jobs.[51] Compared with the costs and benefits provided by formal utility services this is akin to saying that drug dealers and gangs are inner-city entrepreneurs providing useful services such as 'security' and employment opportunities where none exist. I don't want them to lose their livelihoods and they could be redeployed reading meters, collecting bills or working on service extension projects.

Water, oil and gold – a new paradigm or an old cliché?

Giving water a value is an essential part of making sure it is used prudently and managed sustainably, but one can go too far. Twelve years into the current century and the forecasts that water is to be the 'gold' or 'oil' of the twenty-first century do seem to be missing the point. Chris Corbin runs Lotic LLC, a water rights marketing and management company and blogs on these subjects.[52] His '8 reasons why water is not the next gold' post makes a useful case against simplistic desires to see water as a tradable commodity from the perspective of somebody active in water rights.[53] Unlike gold, he argues that water is complex (gold is gold the world over, water has unique characteristics the world over), while gold can be traded and for a known price with many markets and dealers, the value of a water trade depends on time and place, and those rights depend on all manner of things; they are written in water, not engraved in gold. Trading gold is cheap, you get plenty of dollars for every ounce, but trading water is about dealing with high bulk and low value and then there is the cost of shipping it from one place to another. Finally, with water comes regulatory risk and the simple fact that trading water is always going to be a contentious subject.

Willem Buiter's essay 'Water as Seen by an Economist' is part of a note prepared by Citi, the US bank looking at urbanisation and water.[54] The note is a useful summary about various themes, but Mr Buiter's essay is revelatory: 'I expect to see pipeline networks that will exceed the capacity of those for oil and gas today...I see fleets of water tankers and storage facilities that will dwarf those we currently have for oil...I expect to see a globally integrated market for fresh water within 25 to 30 years...water as an asset class will, in my view, become eventually the single most important physical-commodity based asset class.'

While he appreciates that water will have to be sold at different grades (as oil is), Buiter may not have fully appreciated that unlike oil, as John Kemp at Reuters notes, 'for any plausible water price, the value-to-volume ratio would be too low to support an integrated global supply system'.[55] Kemp believes that water per se will never be a global commodity, let alone an asset class. He fully recognises that water must have a value and that its sustainable management depends on it being valued, but that value is determined by local circumstances that the value-to-volume ratio make non fungible. Since oil lies beneath the land and sea and all you need to do is drill a hole, surely it should be free, like water, yet nobody supports such a notion.

Buiter makes points that only a practical economist can, stating that water, rather than being free, 'should be priced or physically rationed to reflect its scarcity value' and even if you allocate some free water at the point of delivery, 'it can never mean that any amount of water is available to the user at zero marginal cost.' When considering water as a 'gift from God' whether one believes in a deity of not, as Buiter notes 'diamonds come free from God' but who is campaigning for diamonds to be free? That comes back to my concern that if water is the gift of God, then sewage is arguably the gift of the Devil. When sewage is free, there is a lot of it around.

Karen Bakker argues that water cannot be a conventional economic commodity because of its relative ubiquity and the unique circumstances of its local availability and when supplied through a network, it is a natural monopoly.[56] In contrast, bottled water is a commodity, being purely something that is collected, bought and sold. However, water can and indeed ought to be regarded as commoditised in an urban context and indeed whenever used by people. Karen's previous book on the politics of water policy was aptly called *An Uncooperative Commodity*.[57]

Investment and development

In a number of countries, most notably China and the USA, state funding for water and wastewater infrastructure projects was increased in 2009 as a response to the recession. The more prosaic aspects of such work, like pipe-laying are pretty labour intensive, so a lot of medium-term jobs can be created for a relatively low outlay, while as we have seen, there is a consistent case that investing in such infrastructure is not a vanity project, but one which delivers attractive economic returns as well as public health and environmental benefits.

The environmental Kuznets curve is a variety of the Kuznets Curve theory, which suggests that inequality increases as a country starts developing and then decreases as development passes a certain point. The environmental Kuznets curve suggests that pollution rises and then falls as a country develops. It is very much open to debate in many areas such as air pollution, but some intriguing relationships in water pollution have been seen. The environmental Kuznets curve kicked in for Japan at $11,000 per capita in 1969 and in Korea in 1987, aided by Japanese technology transfer.[58] The Sceptical Environmentalist Bjorn Lomborg[59] points out how faecal coliform bacteria levels in rivers peak as economies reach $1,375 per capita and bottom out at $11,500 when a second rise starts due to alternative water resources[60] being used. The rise may be due to the data being for 1979 and 1986, when major programmes have seen improvements in river water quality in wealthier countries since 1990.

The message here being that there is a broad understanding that sustainability matters, when people can afford it. That is a somewhat ironic message, as conventional understanding assumes you have to go through an environmental pain barrier before reaching that point. Perhaps more developed countries can lend their expertise to drive down the point at which investing in environmental protection becomes inherently affordable. The assumption (as held by the World Bank in the 1960s) that development could be a tool to allow developed countries to 'export' their pollution ought to be a dangerous anachronism.[61] Singapore is a case in point, where sustainable water management and preventing pollution has become a spur for development.

Sustainability needs sustainable pricing

There is plenty of evidence that appropriate and affordable water tariffs encourage more responsible usage and this is reflected by the relationship between industrial water tariffs and the value created by each unit of water consumed. Revisiting the tariff and affordability data, we can compare it with the overall abstraction of water as a proportion of renewable resources and the amount used by domestic customers.[62]

Tariffs, affordability and domestic water use

Tariff	$ per M3	% of average household income	Domestic use (litres/cap/d)	% of renewable resources abstracted
Mexico	0.49	0.3%	282	16.4%
Japan	1.85	0.3%	345	19.7%
Poland	2.12	1.2%	150	18.3%
Sweden	3.59	0.5%	284	1.5%
France	3.74	0.7%	275	18.2%
Denmark	6.70	0.9%	221	4.2%

This reflects a broader trend that there is little relationship between tariffs and water resources and indeed higher tariffs appear to be charged in water-rich countries and lower ones where water is less abundant. However, where tariffs are a higher proportion of average household income, water usage tends to be lower, most notably in Poland.

By comparing the amount of water used by industry to the amount of money the industrial sector generated in each country, you can get an idea about how much water is valued in each.[63] Countries such as Bulgaria, Kazakhstan, the Kyrgyz Republic and the Ukraine generate less than $1.50 per M³ of water used, suggesting water is cheap and used profligately, while Denmark and Japan get over $100 per M³ of water used, reflecting their advanced industrial base. Countries such as China ($3.67) and Egypt ($2.51) face severe water stress, but get little value from their water while poor countries such as Botswana ($128) and Mozambique ($103) extract all they can. This

336

is an area that has hardly been explored and, indeed, most aspects of water in industrial risk management are just starting to be taken seriously.[64] In other words, when 'every drop counts' companies ought to 'count every drop', but there is little incentive to keep track of finances when water is fundamentally underpriced – even though, to paraphrase Lord Kelvin, 'you cannot manage what you cannot measure.'

Here is where water trading has a value. Not as an asset class for pure speculation, but for encouraging its optimal usage and allocation. The purchase and trading of water rights in the USA and Australia has allowed agricultural users to decide whether some or all of their water would be better off being used by high-value industrial and municipal customers instead of low-grade irrigation. This has been especially effective where farmers and growers have opted to use more efficient irrigation systems and sold their surplus water entitlements. In Australia's Murray-Darling Basin, funds from these sales have also been deployed to allow an Environmental Water Manager to purchase water rights for environmental conservation as well as investing in further efficiency measures.

Investing in water, investing in life

Having inadequate infrastructure is expensive. Wen Jibao, the Chinese Vice Premier noted in 1999 that water scarcity affected the 'survival of the Chinese nation' and the same goes for water pollution.[65] In 2007, the Deputy Minister of the Chinese Environmental Protection Agency estimated that losses from environmental impairment were running at 8–13% of GDP annually.[66] Indeed, 1% of China's GDP is lost through pollution induced water scarcity (RMB 147 million pa) and a further RMB 92 billion through groundwater depletion.[67]

Looking at the costs of not investing in improved water and sanitation, Guy Hutton's Economics of Sanitation Initiative identified health, environmental and other costs such as damage to tourism development that ranged from 1–2% of a nation's wealth in better off South East Asian countries such as Indonesia and the Philippines to 6–7% in poorer countries such as Laos and Cambodia.[68]

Nearly one tenth of the world's disease burden stems from poor water and sanitation.[69] While 10% of all disability adjusted life-years (DALYs) in developing economies stem from water, sanitation and health shortcomings, this proportion is only 0.9% in developed countries. For people living in the former countries, 2.8% of livelihoods are estimated to effectively have been obliterated by disease. Achieving the water and sanitation Millennium Development Goals would avoid 30 million DALYs. Achieving universal access to safe water and sanitation (seen as 98% access) would avoid 585 million DALYs.[70] The gap between the two is remarkable and yet it is almost unremarked. Access to water affects other health impacts, and it is more important than realised.[71] Their 9.1% share of the disease burden related to safe water access works out at 3.6 million deaths and 136 million DALYs each year with more from indirect effects.

In terms of each DALY avoided, the World Bank prefers water and sanitation facilities to cost less than $150, which places a quite low value on a human being's wasted year. For sanitation, it was $270 per DALY, for communal water

$93 and $223 for household water.[72] The difficulty here is that sanitation and sewage treatment are not receiving the attention drinking water provision is getting.[73] This is despite the fact that this is where the real access problems lie. Sanitation investment rates in developing countries are perhaps one quarter of the drinking water level.

Developed countries have been through this and continue to need to provide these services. In the USA, the cost of saving each life by installing effective water treatment (filtration and chlorination) is just $500.[74] That is what revolutionised public health in America's cities a century ago, while over the past 40 years America's wastewater treatment has been transformed through the passing of the 1972 Clean Water Act, putting the focus on rivers management and making modern sewage treatment the norm.[75] The US Environmental Protection Agency believes that the net benefits of the Clean Water Act run at $11 billion per annum or $109 per household.[76]

How urban wastewater treatment developed in the USA, 1940–2008

Million people served	1940	1962	1978	1988	2000	2008
No treatment	32.1	14.7	3.6	1.4	0.2	0.1
Primary	18.4	42.2	44.1	26.5	6.4	3.8
Secondary	20.0	61.5	56.2	78.0	88.2	92.7
Tertiary	0.0	0.0	49.1	65.7	100.9	113.0
No discharge	0.0	0.0	2.2	6.1	14.6	30.0

'No discharge' is where the wastewater is immediately put to beneficial reuse. A recent survey suggests that $188 billion invested in the USA's water and wastewater infrastructure would generate returns of $265 billion and create 1.9 million new jobs both from building this infrastructure and from businesses benefitting from increased activity and opportunities.[77]

Where we stand today is perhaps best summed up by what we do not know. Maria Onestini has found an 'absolute lack of broad-spectrum knowledge on water quality in [Latin America's] poor urban areas' and that this 'points to a lack of acknowledgement of the problem by the science and policy-oriented communities' and so 'the real incidence of disease related to the consumption of unsafe water...by the poor is, by all estimates, much higher than what is reported within the health system(s).'[78]

Realism is needed on cost benefit estimates

The World Health Organization argued in 2004 that $11.3 billion spent each year towards the water and sanitation MDGs would generate a total payback of $84 billion a year.[79] This would come about from improved productivity ($63 billion pa) through freed-up time, the value of deaths averted ($3.6 billion pa), health care savings ($7.3 billion) and the benefits of improved school attendance ($9.9 billion pa). But, as we saw in chapter six, when looking at a study in Morocco, people the world over are people, and given free time, they may elect to enjoy their lives, so I remain rather cautious about reductionist extrapolations of human productivity. A second study published in 2005 estimated that in addition to an eightfold return ($84 billion pa) on funds spent on meeting the water and sanitation MDGs and more intriguingly, a twelvefold return ($344 billion) with universal access to treated water and improved sanitation, but falling to fourfold ($556 billion) with universal household water and sewerage.[80] In each case, the returns are affected by the amount of funding needed.

Studies in the past have tended to assume the complete adoption of a new service, when in reality their use may be seasonal (for example, people using rainwater during wet periods) or incomplete, such as people not sending their septage to a sewage treatment works. This is especially the case in studies on developing economies. It is also clear that the more advanced and costly the intervention, the lower the return will be on that investment.[81] So, septic tanks get a 2.0 to 2.7 times return on their investment against 1.8 times for household sewerage and sewage treatment works had a 1.8 to 2.1 times return. In most businesses, these numbers would be pretty attractive. Indeed, when 38 water-related loans and grants in fifteen African and Asian countries made by the World Bank and the Asian and African Development Banks were examined, the lowest annual return was 5% and the highest 55% – with 36 having returns of 10–40%, good business by any reasonable criteria.

I suspect that claims for powerful financial benefits for investing in water and sanitation have in the past been influenced by the need to monetise everything we do and to look at it in terms of headline-hugging numbers. So my chief reservation about research showing that investment in a certain sector will produce greater returns is that no report would proclaim that investments may not offer good value, just as every business plan presented to potential investors forecasts the new venture thriving in a few years time, while in reality many will fail. Getting carried away about benefits can undermine their credibility, which could be a distraction, especially when there are sound reasons for investing in water and wastewater infrastructure and services. At the local level, the benefits range from freeing people from water collection, to improved health and productivity. At the national level, economic and social development depends on a suitable infrastructure being in place.

Micro-finance needs to avoid mega interest rates

Returns matter – especially if you are paying for water services through micro-credit. The original idea of micro-finance was to provide small loans to those with no access to finance. As it has become a business rather than a calling, these loans are getting costlier.

Microfinance works best if the interest rates are appropriate. Unlike a loan to start a micro-business, the benefits are slower to accrue. One way forward has been shown by the Grameen-Veolia Water joint venture, formed in 2008 between the 2006 Nobel Peace Prize winner and founder of the Grameen microcredit bank Muhammad Yunus and Veolia Water AMI, Veolia Environnement's subsidiary for Africa, the Middle East and India.[82] In 2009 they inaugurated the first water treatment plant intended for village populations living in remote rural areas of Bangladesh. Most of the groundwater in this part of the world is contaminated by arsenic. The plant will supply drinking water to 40,000 people in Goalmari, a village, 100 Km from Dhaka. The water is distributed via a system of storage reservoirs and standpipes. Drinking water is sold to inhabitants for €0.002 per litre, which is 100 times cheaper than locally available bottled water. All profits will be ploughed back to drive the development of other water-related projects in Bangladesh.

Water, money and high finance

While many of the well-established investment banks have a predictable take on water, simply seeing it as a market with needs where spending rises in an effortless straight line, some research titles suggest a more nuanced appreciation is evolving, such as *Peak Water: The Preeminent 21st Century Commodity Story* (Morgan Stanley, 2011),[83] *Water: The Perfect Storm* (Morgan Stanley, 2011),[84] *Water – The Real Liquidity Crisis* (Standard Chartered Bank, 2009),[85] along with *Watching Water – A Guide to Evaluating Corporate Risks in a Thirsty World* (JP Morgan, 2008)[86] and *Water Scarcity: A Bigger Problem Than Assumed* (Merrill Lynch, 2007).[87]

If some of the current crisis' bailed out banks were to make a real contribution to mobilising raising finance for water projects, it might be decent business (direct monetary returns on low coupon debt being modest) while helping to put them, in image terms at least, on the road to redemption.

Plugging the gap – efficiency and propriety

Because water management is not taken seriously enough and because it is too often assumed that spending needs are too high, the scope for making funds go further through efficient management and appropriate technology is often overlooked. We have seen that corruption can eat away 10–40% of funding flows where it is allowed to thrive, and we have also seen in chapter five that innovation offers the possibility of doing more for less. Here are some examples of waste and how it can be remedied.

A World Bank study covering urban Latin America highlights how much of the funding is often already there.[88] Today, tariff revenues are estimated at $19.4 billion against operating costs of $19.1 billion, leaving just $300 million each year for new assets. By ensuring all revenues owing are in fact collected, revenues rise by $2.05 billion to $21.45 billion. By cutting down on overstaffing, $1.82 billion is saved, along with lowering unaccounted for water to 20% a further $1.91 billion can be saved, easing operating costs to $15.37 billion, freeing up $6.08 billion in funding each year to go towards the $12.45 billion needed each year for a proposed $249.2 billion twenty-year plan to make Latin America's water infrastructure globally competitive. The plan is not perfect, envisaging some slums remaining unconnected and 64% of sewage properly treated, but it is well in advance of any other plan on the table.

The scale of inefficiency where it matters most was highlighted when a number of Sub-Saharan countries were subjected to an AICD (Africa Infrastructure Country Diagnostic) test by the World Bank.[89] Looking at water, they found that the percentage of a country's GDP wasted by mispricing, unaccounted-for water and collection inefficiencies ranged from 0.2–0.3% (Nigeria and Uganda) to 1.2% (Ghana). By working at inefficiency and the clearest mispricing, the World Bank believes that 32% of the funding gap needed for proper water and sanitation services can be eliminated.[90]

Italy is a depressing example of money misspent, with water and wastewater projects in Italy's south making a minimal impact. In Sicily, the EU provided €8.5 billion in funds under its Agenda 2000 programme to modernise the island's infrastructure. €700 million was spent on water supply, during which time the percentage of the population who received only intermittent water rose from 33% to 39%. Likewise, €400 million spent on upgrading water treatment saw an increase in households accessing treated water from 43% to 47% or €2,000 per person receiving treated water. At a total of €221 investment in water per person over the whole island (Sicily's population was 4.97 million in 2001) the proportion of these funds actually being spent on water projects must have been small.

In Europe, using the private sector to develop sewage treatment assets has driven down capital costs by 15–40% since the early 1900s. The private sector has a broad remit for driving down costs. Small bore sewerage networks laid down with local labour in El Alto, Bolivia between 2000 and 2002 under a concession managed by Suez cost $90 per capita, 40% of previous fees and more generally, the use of free local labour can drive down the cost of sewerage and water connections by up to 40%.[91]

Economies of scale matter.[92] For example, the unit construction cost of traditional filter type water treatment plants serving 50,000 people cost 15% of the amount per person served as do those serving 500, while the unit cost of a water pumping station serving 50,000 people cost 19% of the per person cost as those serving 500. Although decentralisation is generally seen as a good thing in utilities, it can impair economies of scale, especially when it comes to the duplication of tasks and capabilities.[93] Smaller utilities are especially vulnerable when it comes to their purchasing power for goods and services and their own professional capabilities. They may also be more entangled in the varying obligations and desires of local governments. Even so, nothing is quite as simple as it appears, so that apparent economies of scale can be something of a chimera, since large cities are typically served by large utilities and large cities have a higher population density than towns and villages meaning that connections are closer together. Also, the bigger the city, the more challenging it is to ensure sustainable water supplies and large sewage treatment works can be self-defeating because of the cost of developing and operating major sewer mains systems. As a result, while a study in Germany found the optimal utility size to be 66,000 people, it was 766,000 people in Japan.

Another area to consider is how much money is lost through leakage. Defining leakage and identifying distribution losses and non-revenue water is an evolving subject but it is clear a lot of valuable water is getting lost. A recent study focussing on Asia identified 21.4 billion M^3 getting lost through leaks and 7.3 billion M^3 through unpaid bills. At their assumption of $0.30 per M^3 this worked out as an annual loss of $8.6 billion. Even halving this would save $172 billion over 40 years, a useful part of their funding needs.[94]

Countering corruption

Corruption by its very nature defies precise analysis, but orders of magnitude can be telling. A study of illicit financial flows in developing countries between 2000 and 2009 estimated that 'kickbacks, bribes, embezzlement, and other forms of official corruption' in developing economies cost $383 billion per annum.[95] In India Transparency International[96] estimates the cost of petty corruption when dealing with eleven arms of the government costs Rs 211 billion per annum ($4.7 billion), while a UN survey[97] of 5,000 people put the cost at Rs 268 billion ($5.6 billion) each year where the public have to pay various public departments to carry out what they are meant to do in the first place. If this $5 billion per annum in petty corruption were diverted towards water and sewerage service extension, urban coverage would be a reality within two decades. A real basis for sustainable development would be built and the Ganges would return to life.

Italy is a depressing example of money getting misappropriated, with water and wastewater projects in the south making a minimal impact. In Sicily, the EU provided €8.5 billion in funds under its Agenda 2000 programme to modernise the island's infrastructure.[98] €700 million was spent on water supply, during which time the percentage of the population who received only intermittent water actually rose from 33% to 39%. Likewise, €400 million spent on upgrading water treatment saw an increase in households accessing treated water from 43% to 47% or €2,000 per person receiving treated water. At a total of €221 investment in water per person over the whole island (Sicily's population was 4.97 million in 2001) the proportion of these funds actually being spent on water projects must have been small.

Corruption can be countered. Cost recovery reduces the scope for hidden subsidies and payments. It has to be linked with proper financial and service delivery reporting as when people pay for a service, they are stakeholders as well. Independent regulation is needed (seen in less than 25% of developing countries by 2004) along with avenues for consultation and complaint by user communities. Direct payment of bills to offices in person, post or via mobile rather than to agents removes a layer of corruption.

Windows of transparency work when they expose corrupt practices.[99] Before 2005, pipe manufacturers in Colombia had to pay bribes worth 12% of the contract value to get an order. Prices fell after the association of pipe manufacturers entered into an anti-corruption agreement with Transparency International. Likewise, an integrity pact developed by TI saw costs for the Greater Karachi Water Supply Scheme's Phase V Stage II fall 18% below initial expectations. Water integrity national surveys (WINS) can be developed for national anti-corruption plans and used by local utilities. But old habits need to be confronted. A survey of 59 countries found that in nearly half, audited information was withheld for at least two years, and in ten countries information was not even made available to legislators. Somebody has something to hide.

Good governance in Africa

Are even decent water services affordable and doable in the less developed countries? I believe so. Uganda's National Water and Sewerage Corporation was a classic 'basket case' utility, water running at 10% of capacity and only for two hours a day, making vendors the norm. In 1987, the utility was reformed and improved its service coverage, but it was funded by government debt at commercial interest rates. In November 1998 William Muhairwe took over as Managing Director with a mandate to make the utility pay for itself. Something was going right, as the Corporation had grown its revenues fourfold to 21 billion Shillings even with five billion Shillings of bill arrears every year, but the Corporation was losing 4.2 billion Shillings due to its debts.[100] Following seven phases of work under Muhairwe, the Corporation has been transformed and as these took effect, the Government converted its 153 billion Shilling debt into equity freeing the company from its debt service costs.[101]

The main focus has been to increase revenues by more connections and making informal providers unnecessary. Between 1998 and 2011, the number of staff per 1,000 connections fell from 36 to six, while connections rose from 50,826 to 272,406, meaning that even in a period of unprecedented population growth and urbanisation, coverage increased from 48% to 75%. With all connections metered, non-revenue water fell from 60% to 33% and bill paying up from 60% to 96% revenues rose to 113 billion Shillings.[102] To encourage connections, customers living within 50 metres from the main water pipe are connected free of charge, along with paying their monthly water bills directly to the Corporation.[103] A connection fee of 50,000 Shillings is paid by those living further from the pipe. Water tariffs are 1,519 Shillings per M^3 for domestic customers and 982 when from public standpipes[104] compared with about 5,000–25,000 for water from a vendor.[105] Has this worked? Instead of a two billion Shilling loss in 1998, there was a 9.4 billion Shilling profit in 2010. With confidence in the system and the service, sewerage is now being developed, with 6.5% of the population having household sewerage in 2010, along with formal wastewater treatment.[106]

It is tempting to always say that the answer for Full Cost Recovery is to raise the tariff to a sustainable level. Sometimes, this misses the real point, which is that the tariff is burdened with unnecessary costs ranging from over-staffing, corruption and mismanagement to along with improving collection efficiency so that the money raised goes to where it matters.[107]

Conclusions – connecting everybody is not cheap but excellent value

Your debts mount high – ye plunge in deeper waste
    James Smith and Herbert Smith, *Rejected Addresses*

We have seen that poor service and coverage are the norm where water finance is not taken seriously. In economic terms, water is a natural monopoly, a service that cannot be competitively replicated due to the high cost of developing a duplicate infrastructure. There is indeed a monopoly here, but sometimes it is one of water policy setting and analysis by those who are perhaps more interested in the theory of natural monopolies than the practical side of regulating them. Comprehensive coverage is a part of coherent and sustainable water management policy, but getting policymakers to appreciate this remains a challenge. Much of what is gained from safe water and sanitation access is improved health, which can be quantified. However, lifestyle and environmental benefits are by their nature externalities and cannot be quantified and so these benefits cannot be directly included in a project's achievements.

As a result, willingness to pay remains a problem, not the least when tariffs rise before water services are improved. The unserved are willing to pay an appropriate fee, but it must be affordable and the service must be delivered. Affordability lies at the heart of any financing strategy, whether using Full Cost Recovery or Sustainable Cost Recovery. Other tariff options are rarely capable of delivering sustainable services. The argument that water is 'a gift of God' and should be free omits 'those other free gifts from God: cryptosporidiosis, dysentery and cholera' which were conveniently overlooked in 2011 when Italy voted against the principle of Full Cost Recovery in a referendum.[108]

Water will remain an 'uncooperative commodity' but its inherent value has to be recognised. There ought to be a distinction between those who want water properly valued and traded to encourage infrastructure funding and to ensure its efficient allocation and management against those who seek to trade water rights in pure acts of asset class speculation. The latter are manna for the anti-finance campaigners and provide cheap headlines that can do as much harm as those who proclaim that all water should be free.

As with those who are over-keen to 'invest' in water, there are others who remain indifferent to its funding needs. We have to understand that aid will not get better, or more to the point, it will not be better to the point where it will be able to significantly dent the funding gap. Aid always helps, it makes things happen faster and better if well used, but it will not be the answer in developing economies, as it is best used as a catalyst for other funding. The NGOs such as WaterAid, Pump Aid, Tearfund and Cafod are something else and nothing must distract them from their task, as they are amongst the most effective and accountable of the organisations delivering on-the-ground benefits. They matter, especially at a local level; since 1981 WaterAid has provided water to 16 million and since 2001, sanitation to 11 million.

Looking ahead, all being well, the world's population will stabilise at some point towards the end of this century before easing to a more sustainable level in the longer term. Likewise, the potential for a new equilibrium between urban and rural populations exists. But, as we have seen by the shifts in urban and rural populations in the UK, technological and social change continue to make the interplay between urban and rural populations a dynamic one. Before such equilibriums will be reached, two more generations will live their lives. Must they endure the policy and physical disconnects of today? Coherent, properly phased connection policies can deliver universal and affordable services.

Without a fundamental change in attitudes and priorities, a lack of water and sanitation at various levels will continue to be the urban norm. Without a vision for universal water and sanitation access it is doubtful that this will happen before 2100. This is unsustainable both in terms of human decency and sustaining life on earth. This is more than a decision about what we want to spend ourselves, it also affects the way our children and grandchildren will live and how they view our generation.

Planning and prevention is better than a cure, and looking hard at the cost of projects and comparing what supply and demand management can deliver offers the prospect of avoiding spending on new assets that could have been avoided. Singapore's infrastructure was not built in a day, it developed along with the country's economy, spurring the country's development rather than trailing in its wake.

Full Cost Recovery is the ideal, as it allows a utility to be financially self-sufficient. Where affordability issues are an impediment, Sustainable Cost Recovery offers a financially viable alternative. In each case, it is about ensuring the funding is in place to allow for Total Expenditure (TOTEX, or CAPEX and OPEX combined) to deliver full access to appropriate water and sewerage services. By driving out corruption, optimising the efficiency of operations, cost-effective procurement of hardware and smart demand management, these costs can be driven down, making the services more affordable to all.

1   Foster V & Yeps T (2005) Is Cost Recovery A Feasible Objective for Water and Electricity? A study quoted in Fay M & Morrison M (2005) Infrastructure Development in Latin America and the Caribbean: Recent Developments and Key Challenges, Report 32640-LCR, World Bank, Washington DC, USA

2   OECD (2010) *Pricing Water Resources and Water and Sanitation Services*, OECD, Paris, France

3   European Union (2000) Directive 2000/60/EC of the European Parliament and of the Council establishing a framework for the Community action in the field of water policy, *Official Journal* (OJ L 327) on 22 December 2000

4   Winpenny J (2003) *Financing Water For All: Report of the World Panel on Financing Water Infrastructure,* World Water Council

5   Rees C J, Winpenny J & Hall A (2008) Water Financing and Governance, Global Water Partnership TEC Background Paper 12

6   Owen D A Ll (2009) *Tapping Liquidity: Financing water and wastewater to 2029*, Thomson Reuters, London, UK

7   Foster V & Yeps T (2005) Is Cost Recovery A Feasible Objective for Water and Electricity? A study quoted in Fay M & Morrison M (2005) Infrastructure Development in Latin America and the Caribbean: Recent Developments and Key Challenges. Report 32640-LCR, World Bank, Washington DC, USA

8   OECD (2009) *Managing water for all: An OECD perspective on pricing and financing,* OECD, Paris, France

9   OECD (2010) *Pricing Water Resources and Water and Sanitation Services*, OECD, Paris, France

10 Smets H (2008) Water for domestic uses at an affordable price, presentation to the International Conference on the Right to Water and Sanitation in Theory and Practice, Oslo, Norway

11 Klawitter S (2008), Full Cost Recovery, Affordability, Subsidies and the Poor – Zambian Experience, presentation to the International Conference on the Right to Water and Sanitation in Theory and Practice, Oslo Norway

12 Cosgrove W J & Rijberman F R (2000) *World Water Vision: Making Water Everybody's Business,* Earthscan, London, UK

13 Vision 21 (2002) Vision 21, World Water Council
    Briscoe J (1999) The changing face of water infrastructure financing in developing countries, *International Journal of Water Resources Development*, 15 (3), 301–308
    GWPFA (2000) Global Water Partnership, Towards Water Security: A Framework for Action

14 Environmental Resources management (2003) *The European Union Water Initiative: Final Report of the Financial Component,* ERM, London, 2003

15 Lenton R & Wright A (2004) Interim Report of Task Force 7 on Water and Sanitation, Millennium Project, United Nations, New York, USA

[16] Hutton G & Bartram J (2008) Global costs of attaining the Millennium Development Goal for water supply and sanitation, Bulletin of the World Health Organization 2008; 86:13–19.

[17] Prynn P & Sunman H (200) Getting the water to where it is needed and getting the tariff right, FT Energy Conference, Dublin 11–2000

[18] Newing R (2011) Water: smart ways to reduce stress levels, *Financial Times*, 28th October 2011.
http://www.ft.com/cms/s/0/e75a17f0-fbfa-11e0-989c-00144feab49a.html#axzz1cLx3yHQz

[19] Booz Allen Hamilton (2007) *Lights! Water! Motion!*

[20] Heymann E et al. (2011) Water: Investments of €500 bn required – every year!, Deutsche Bank Research. Research Briefing, June 6, 2011
Heymann E et al. (2010) High investment requirement mixed with institutional risk, Deutsche Bank Research. Current Issues, June 1, 2010

[21] GWM 2011 (2010) *Global Water Markets 2011,* Global Water Intelligence, Media Analytics Limited, UK

[22] Owen D A Ll (2006) *Financing water and wastewater to 2025: From necessity to sustainability*, Thomson Financial, London, UK
Owen D A Ll (2009) *Tapping Liquidity: Financing water and wastewater to 2029,* Thomson Reuters, London, UK

[23] GWI (2011) France Faces up to new EU compliance demands & Private finance rises up Spain's water agenda, *Global Water Intelligence*, 12/10, October 2011

[24] Danish EPA (2001) The Environmental Challenge of EU Enlargement in C&EE, DANCEE – Danish Cooperation for Environment in Eastern Europe, Ministry of the Environment, Copenhagen, Denmark
EEA (2005) Report 2/2005 Effectiveness of urban wastewater treatment policies in selected countries, EEA, Copenhagen, Denmark

[25] Burnside & Associates (2005) Water & Wastewater Asset Cost Study, Ministry of Public Infrastructure Renewal, Ontario, Canada
FEASIBLE model (Dancee, 2004)
UNDP (2005) Water and Sanitation Needs Assessment Model User Guide, UN Millennium Project (www.unmillenniumproject.org)

[26] UNDP (2005) Water and Sanitation Needs Assessment Model User Guide, UN Millennium Project (www.unmillenniumproject.org)
WHO, Global Water Supply and Sanitation Assessment 2000 and WHO (2008) Regional & Global Costs of Attaining the Water Supply and Sanitation Target of the MDGs, WHO, Geneva, Switzerland
Hutton G & Bartram J (2008), Global costs of attaining the Millennium Development Goal for water supply and sanitation, Bulletin of the World Health Organization, 2008; 86: 13–19
ISW (2005) Blue book Mali, ISW, Montreal, Canada

ISW (2005) Blue book Burkina Faso, ISW, Montreal, Canada

ISW (2005) Blue book Niger, ISW, Montreal, Canada

WaterAid (2004) The Water & Sanitation MDGs in Nepal, WaterAid, Nepal

WaterAid (2005) $2 billion dollars, the cost of water and sanitation MDGs for Tanzania, WaterAid, UK

UN (2005) UN Millennium Project Task Force on Water & Sanitation, 2005

[27] Transparency International (2008) *Global Corruption Report 2008: Corruption in the Water Sector,* Cambridge University Press, Cambridge

[28] GWI (2007) *Global Water Markets 2008,* Media Analytics Limited, Oxford, UK

[29] Owen D A Ll (2009) *Tapping Liquidity: Financing water and wastewater to 2029,* Thomson Reuters, London, UK

[30] US EPA (2002) *The Clean Water and Drinking Water Infrastructure Gap Analysis,* US EPA, Washington DC, USA

[31] US EPA (2010) *Clean Watersheds Survey, 2008 Report to Congress,* US EPA, Washington DC, USA

[32] US EPA (2009) *Drinking Water Infrastructure Needs Survey and Assessment, Fourth Report to Congress,* USEPA, Washington DC, USA

[33] Fitch IBCA (2000) Secure Credit on Tap, Fitch Ratings, 7th July 2000

[34] www.tremolet.com

[35] Tremolet S (2010) *Innovative Financing Mechanisms for the Water Sector*, OECD, Paris, France

[36] CWSRF (2009) Annual Report 2008, US EPA

DWSRF (2008) Annual Report 2007, US EPA

[37] Kowaco (2009) 2008 Annual Report, Kowaco, Seoul, Korea

[38] International Financial Review / Thomson Reuters database

[39] USAID (2004) Effective Catalysts for Private Investment in Municipal Finance, presentation to the CDA Conference, Washington DC, 1 October 2004

[40] Tremolet S (2009) Private money for public water – A safe haven in the midst of a financial storm?, A presentation to ICEA, 16 June 2009

[41] Tremolet S (2009) Private money for public water – A safe haven in the midst of a financial storm?, A presentation to ICEA, 16 June 2009

[42] Colombia Infrastructure Group LLC, Press Release, 15th December 2009

[43] Iff-immunisation.org

[44] OECD (2009) Measuring aid to water and sanitation, OECD-DAC, Paris, France

[45] Data from the OECD's QWIDS Query Wizard for International Development Statistics, Stats.orcd.org/qwids

46 Tremolet S & Evans B (2010) Output-Based Aid for Sustainable Sanitation. OBA Working Paper Series No 10, September 2010, GPOBA, World Bank, Washington DC, USA

47 Tremolet S, Kolsky P & Perez E (2010) Financing On-Site Sanitation for the Poor: A Six Country Comparative Review and Analysis, Water and Sanitation Program: Technical paper, WSP, World Bank, Washington DC, USA

48 Bakker K (2010) *Privatizing Water: Governance Failure and the World's Urban Water Crisis,* Cornell University Press, USA

49 World Bank's (2011) PPIAF 2011 Water & Sanitation Sector Review, Data is adapted from http://ppi.worldbank.org/explore/ppi_exploreSector.aspx?sectorID =4.

50 World Bank (2005) Water Supply and Sanitation Lending: Volume Rises, Quality Remains High, Water Supply and Sanitation Feature Stories, Washington DC, USA

51 Solo T M (1999) Small-scale entrepreneurs in the urban water and sanitation market, *Environment and Urbanization*, April 1999 vol. 11 no. 1 117–132

52 www.activelymovingwater.com

53 http://activelymovingwater.com/gold/8-reasons-why-water-is-not-the-next-gold/ posted 18th October 2010

54 Citi (2011) *Global Themes Strategy: Thirsty Cities – Urbanisation to Drive Water Demand,* Citicorp Global Management, New York, USA

55 http://www.commodities-now.com/reports/environmental-markets/7138-water-is-not-the-next-big-asset-class, accessed 1st September 2011

56 Bakker K (2010) *Privatizing Water: Governance Failure and the World's Urban Water Crisis,* Cornell University Press, USA

57 Bakker K (2003) *An Uncooperative Commodity: Privatizing Water in England and Wales,* Oxford University Press, Oxford, UK

58 Takahashi K (2009) In Search of a Comprehensive Approach to Sustainable Management of Water Resources in the World Community, in Biswas A K, Tortajada C & Izqueirdo (eds) *Water Management in 2020 and Beyond,* Water Resources Development and Management, Springer-Verlag, Berlin, Germany

59 Lomborg B (2001) *The sceptical environmentalist*, Cambridge University Press, Cambridge, UK

60 Shafik N (1994) Economic development and environmental quality: an econometric analysis, Oxford Economic Papers, 46: 757–773 WDR (2002) *World Development Report 2002: Development and the Environment*, World Bank, Washington DC, USA

61 Bakker K (2010) *Privatizing Water: Governance Failure and the World's Urban Water Crisis*, Cornell University Press, USA

62 OECD (2008) *OECD Environmental data Compendium 2008*, OECD, Paris, France

63 WWDR 2 (2006) *World Water Development Report 2: Water: A*

Shared Responsibility, UNESCO, Paris, France

[64] CDP (2011) CDP Water Disclosure Global Report 2011: Raïsing corporate awareness of global water issues, Carbon Disclosure Project, London, UK

[65] *Economist* (2005) Drying up, *Economist*, 19 May 2005, p46

[66] *Economist* (2007) A Ravenous Dragon: Special report on China's quest for resources, *Economist*, 5 March 2007, p18

[67] World Bank (2009) Cost of Pollution in China: Economic estimates of physical damages, World Bank, Washington DC, USA

[68] Hutton G (2010) The Costs and Benefits of Water & Sanitation Services: Preliminary Results of the Economics of Sanitation Initiative in Southeast Asia, Paper at the OECD Expert Meeting on Water Economics and Financing, OECD, 17th March 2010, Paris, France

[69] Prüss-Üstün A, Bos R, Gore F & Bartram J (2008) *Safer water, better health: costs, benefits and sustainability of interventions to protect and promote health*, World Health Organization, Geneva, Switzerland

[70] World Health Organization (2002) *World Health Report 2002,* WHO, Geneva, Switzerland

[71] Prüss-Üstün A (2008) *Safer Water, Safer Health. Cost, Benefits and Sustainability of Interventions to Protect and Promote Health*, WHO, Geneva, Switzerland

[72] Cairncross S & Valdemanis V (2006) Water Supply, Sanitation and Hygiene Promotion, in *Disease Control Priorities in Developing Countries*, Washington DC, USA

[73] Tremolet S (2011) *Benefits of Investing in Water and Sanitation*, OECD, Paris, France

[74] Cutler D and Miller G (2005) The Role of Public Health Improvements in Health Advances: The Twentieth-Century United States. Demography, 42: 1–22

[75] US EPA (2010) *Clean Watersheds Survey. 2008 Report to Congress*, U EPA, Washington DC, USA

[76] Bingham T H et al (2000) A Benefits Assessment of Water Pollution Control Programs Since 1972: Part 1, The Benefits of Point Source Controls for Conventional Pollutants in Rivers and streams, US EPA, Washington DC, USA

[77] Gordon E et al (2011) *Water Works: Rebuilding Infrastructure, Creating Jobs, Greening the Environment,* Green For All, USA

[78] Onestini M (2011) Water Quality and Health in Poor Urban Areas of Latin America, *International Journal of Water Resources Development*, 27, 1, 219–226

[79] Hutton G & Haller L (2004) Evaluation of the costs and benefits of water and sanitation improvements at the global level, World Health Organization, Geneva, Switzerland.

[80] WHO (2005) Making water part of economic development: The economic benefits of improved water management and services,

Stockholm International Water Institute and World Health Organization, Stockholm, Sweden

[81] Hutton G (2010) The Costs and Benefits of Water & Sanitation Services: Preliminary Results of the Economics of Sanitation Initiative in Southeast Asia, Paper at the OECD Expert Meeting on Water Economics and Financing. OECD, 17th March 2010, Paris, France

[82] Veolia Environnement (2009) VE Press Release, 24 June 2009

[83] Morgan Stanley (2011) *Peak Water: The Preeminent 21st Century Commodity Story*, Morgan Stanley Smith Barney, NY, USA

[84] Morgan Stanley (2010) *Water: The Perfect Storm*, Morgan Stanley Smith Barney, NY, USA

[85] Barrett A, Glendenning C & Lundgren (2009) *Water – the real liquidity crisis*, Standard Chartered Bank, London, UK

[86] Levinson M (2008) *Watching water – A guide to evaluating corporate risks in a thirsty world*, JP Morgan Chase & Co, New York, USA

[87] Knight Z & Miller-Bakewell R (2007) *Water scarcity; a bigger problem than assumed*, Merrill Lynch Research, London, UK

[88] Mejia A (2011) Responding to Global Changes: Water in an Urbanizing World. Challenges for Urban Water Provision in Latin America and the Caribbean, Presentation given to the Stockholm World Water Week, Stockholm, Sweden, 24th August 2011

[89] Briceno S, Smits K & Foster V (2008) Financing Public Infrastructure in Sub-Saharan Africa: Patterns, Issues and Options, AICD Paper 15, World Bank, Washington DC, USA

[90] Briceno S & Foster V (2009) Africa's Infrastructure: A Time for Transformation, Presentation by AICD for the World Bank Water Week, February 2009, Washington DC, USA

[91] WHO / UNCF / WSSCC (2000), *Global water supply and sanitation assessment report*, Geneva and New York
Hutton G & Bartram J (2008), Global costs of attaining the Millennium Development Goal for water supply and sanitation, Bulletin of the World Health Organization, 2008; 86: 13–19

[92] US EPA (2001) *1999 Drinking Water Infrastructure Needs Survey: Modelling the Cost of Infrastructure*, US EPA Office of Water, Washington DC, USA

[93] OECD (2009) *Report on measures to cope with over-fragmentation in the water supply and sanitation sector,* Report by Kommunalkredit Public Consulting, OECD, Paris, France

[94] Frauendorfer R & Liemburger R (2010) *The Issues and Challenges of Reducing Non-Revenue Water*, ADB, Manila, Philippines

[95] Kar D & Curico K (2011) Illicit Financial Flows from Developing Countries: 2000–2009, Global Financial Integrity, Washington DC, USA

[96] Centre for Media Studies (2005) India Corruption Study 2005, Transparency International India, New Delhi, India

359

[97] http://unpan1.un.org/intradoc/groups/public/documents/APCITY/UNPAN019888.pdf

[98] La Repubblica, (2010) Fondi europei, spesa fallimentare. 230 milioni per 8 chilometri di ferrovia, 24 September 2010, http://palermo.repubblica.it/cronaca/2010/09/24/news/agenda_2000_il_conto_finale_spesi_8_5_miliardi_senza_risultati-7374077/

Presseurop (2010) Sicily, gobbler of EU funds 8 October 2010 (translation of the original article appeared in *La Stampa*) http://www.presseurop.eu/en/content/article/356171-sicily-gobbler-eu-funds

[99] Transparency International (2008) Global Corruption Report 2008: Corruption in the Water Sector, CUP, Cambridge, United Kingdom

[100] Muhairwe W T (2002) Strategies and challenges of improving water and sanitation service delivery – a case of National Water and Sewerage Corporation, Uganda, Water 21, Africa Energy Forum, London, July 2002.

[101] Muhairwe W T (2011) The Turnaround Account: Reforming the National Water and Sewerage Corporation. Address to the OECD Global Forum on the Environment, OECD, Paris, France, 26th October 2011

[102] NSWC (2010) National Water and Sewerage Corporation, Annual report 2009-2010, NSWC, Kampala, Uganda

[103] Muhairwe W T (2005) Performance Improvement Programmes, the case of NWSC – Uganda, Managing Water Supply and Sanitation in Large Cities and Urban Areas WSP Workshop, 23–24th February 2005, Karachi, Pakistan

[104] Muhairwe W T (2009) Fostering Improved Performance through Internal Contractualisation, presentation to the 5th World Water Forum, Istanbul, Turkey, March 2009

[105] Pangare G & Pangare V (2008) Informal Water Vendors and Service Providers in Uganda: The Ground Reality, The Water Dialogues www.waterdialogues.org, accessed 28th October 2011

[106] NSWC (2010) National Water and Sewerage Corporation, Annual report 2009-2010, NSWC, Kampala, Uganda

[107] Taweel M Z (2011) Cost Recovery and Mismanagement Costs At Water & Wastewater Utilities In the Developing World, Presentation given to the Stockholm World Water Week, Stockholm, Sweden, 24th August 2011

[108] Gasson C (2011) Italy's grasshoppers presage a bleak winter for water, *Global Water Intelligence*, 12, 6, June 2011

Chapter 9

Private Vice, Public Virtue?

The F-word is finance: Kyoto, May 2003

The moment he saw the media representatives gathered at the front of the hall, Jim Winpenny felt apprehensive.[1] Various sessions covering the relationship between water and youth, gender and culture at the Third World Water Forum had run smoothly and now it was financing water's turn. Jim was the rapporteur for the session, chaired by Michel Camdessus, the author of *The Camdessus Report* that had earlier that year sought to consider how much needs to be spent to provide wider access to water and sanitation worldwide and how this could be financed.[2] A few minutes into Monsieur Camdessus's opening remarks, the fire exits were flung open and the cameras trained on a group of demonstrators who physically broke the meeting up. One demonstrator told Camdessus that an important issue was only mentioned in the footnotes and Camdessus replied 'but I always read the footnotes first in any document!'[3] Two world views in mutual incomprehension and financing water would not be discussed at that forum.

Three years earlier, at the Second World Water Forum in The Hague, a 'World Water Vision' was unveiled. It set out to attain universal access to water, sanitation and urban wastewater by 2025. Here one session was disrupted by a pert pair of naked protestors with anti-private sector slogans on their buttocks and it took the diplomatic skills of Willem-Alexander, Prince of Orange and patron of the Global Water Partnership, to restore order. The World Water Forum is meant to be a triennial get-together for policymakers and thinkers to consider water and sanitation concerns, but the NGOs were not in a mood for listening. The World Water Vision was quietly shelved in the wake of market uncertainty as the first dot-com boom turned to bust. Campaigner hostility to the idea of people paying for water made abandoning these aims much simpler. A decade on, there has been no concerted drive towards universal water and sanitation provision. How did we get to the point where the practicalities of financing and managing water became taboo subjects?

At 2006's WWF4 in Mexico, the finance session was interrupted by chants of 'water is a right, not a commodity' and then allowed to continue. There were no disruptions at Istanbul's WWF5 in 2009.[4] Indeed, there was a plethora of papers and presentations (at least to a data junkie like me), but the Forum lacked the focus needed for a memorable message to emerge.

In the preceding chapters it has been noted that one of humanity's greatest shortfalls lies in its collective inability to provide adequate access to water and sanitation. It is also evident that a powerful lobby exists to sustain this unsustainable status quo because such shortfalls should not be challenged if they threaten the free and anti-private sector water shibboleths.

The right to water and the private sector

The United Nations has seen its fair share of skirmishes over the ever evolving definition of the Human Right to Water.[5] Today, this right means a regular supply of safe drinking water sufficient for domestic needs and sanitation that is safe for people and the environment, affordable and easily accessible. It would be hard to disagree with this and it may in time become a tool for obliging governments to serve the unserved. When this was debated up in 2010, an intense lobbying process was mounted to make it the 'right to free water' and to explicitly ban any form of private service provision, presumably extending to the street vendors as well. Pragmatism prevailed, and the lobbying did not get far. Indeed, when the paper emerged it was sensible and unobjectionable, being passed by 124 nations to nil. The paper notes that 'businesses and the private sector are important players, too' and that 'the right to water requires water services to be affordable for all...the human rights framework does not provide for a right to free water',[6] while the UN Secretary-General Ban Ki-Moon observed that 'there is no longer any doubt that business plays an integral role in delivering economic and social progress... provided access to safe water and sanitation and advanced environmental sustainability.'[7]

What is private sector participation and why do we have it?

Involving the private sector is seen as perfectly normal in many walks of life from growing and selling food to utilities like telecoms. In water, it can be a contentious subject at the best of times. Private Sector Participation (PSP, sometimes called PPP or Public-Private Partnership) is when private operators are involved in the management of water and wastewater services. One of the ironies in this debate is that the private sector is responsible for developing much of the world's water infrastructure irrespective of who runs it – to such extent that in 1971, at the height of Mao's Cultural Revolution, China was ordering water treatment hardware from France's Degrémont. Here we ought to make a distinction between PSP and privatisation, since the latter really refers to the outright ownership of the assets as well as managing them, as has been the case in England and Wales since 1989. I regard PSP as an answer, but not the whole answer. It is one of a quiver full of arrows that sustainable water management needs. The private sector may be used where it may mobilise funding, improve the efficiency of operations, bring about innovation or depoliticise water management by distancing it from local and national politics.

There is a preliminary level of engagement and three actual degrees of private sector management and involvement:

Commercialisation

Commercialisation calls for the municipal water and/or sewerage entity to be operated on its own without hidden subsidies. This has been widely used either as an end into itself or as a prelude to more extensive private sector participation. Madrid's Canal Isabel II has been run like this since 1853, with private sector plans only emerging in the wake of the current financial crisis. As we have seen, service provision in Uganda and Cambodia's Phnom Penh benefitted from this approach. To blur matters, some state-run water companies such as Vitens of the Netherlands and Aguas de Portugal have taken on projects in developing economies. NGOs campaigned against them seeking these contracts, even though it could be argued that a private project managed by a state-owned entity could be 'better' than a private company.

O&M and lease contracts

The entry level for PSP is Operations & Maintenance (O&M) or Lease contracts for a site or a municipality's services. O&M contracts usually operate on a fixed-fee basis for a fixed number of years running a facility on behalf of a municipality. Lease contracts typically manage the assets and billing, but not capital expenditure. The French 'Affermarge' is a lease contract, traditionally running for seven years and then put out to tender for renewal. These two types of contract do not delegate full financial responsibility to the private operator, especially with regards to capital investments.

Concessions

Concessions involve private operation of assets to pay for new facilities or upgrading work. Build-own-operate (BOO) and build-operate-transfer (BOT) contracts relate to a specific programme of capital improvements, while the full utility concession contract embraces all aspects of service provision and capital spending. An independent regulator is usually needed because of the risks involved for all stakeholders.

A BOO/BOT project's cash flow is usually contractually pre-determined, often with government backing. There is an element of construction risk, but the absence of market risk means that the project can have more debt loaded in than in a concession. A project's construction risk can also be mitigated, whereby a facility already generating cash flow gets taken over for expansion by the private sector. Therefore BOT/BOO projects are an effective means of rapidly organising private capital and management towards a narrow range of services. As they do not affect the utility's management and operation, underlying problems such as leakage (and illegal interception), over-staffing and poor tariff collection may not be addressed and a poorly framed BOO/BOT contract may in fact delay system-wide improvements. In full utility concessions, existing operations are taken over and revenues can be used immediately minimising construction risk. A more robust balance sheet can be created over time, allowing the prospect of selling long-term debt through bonds.

A hybrid privatisation has emerged from a number of these commercial entities where the municipality floats some of the shares of the entity while retaining majority ownership and therefore overall governance. The best example is Sao Paulo's SABESP in Brazil, whose shares and bonds are actively traded on the Sao Paulo and New York exchanges, with the municipality holding 53% of the company's equity. Other examples are to be found elsewhere in Brazil along with China, Italy and Greece.

Privatised – asset owning

The most dramatic and politically contentious form of privatisation is the outright sale of the utility's assets. To date this has been used in the 1989 sale of the English and Welsh water and sewage companies (WaSCs), along with two examples in the Czech Republic and some in China and Chile. While the assets are in private hands, the licence to operate them may be subject to renewal. In the USA, the Water Only Companies of England and Wales and in one case in India (Tata's Jamshedpur City) assets were developed from the outset by the private sector and they continue to be operated by them.

The greater the degree of private involvement the more political risk it carries and when assets go into private hands, there is a danger of losing the customers' sense of civic duty. During the 1976 drought, water consumption in England and Wales fell and standpipes and supplies brought in by tankers were accepted stoically. 'Share a bath with a friend,' suggested Dennis Howell, the then Minister for Drought. In contrast, during the 1995 drought, consumption rose amidst intense bitterness even at the possibility of water restrictions being imposed. They were not, but public goodwill was unintentionally divested as well.

O+M and concession contracts are sometimes referred to as the 'French Model', while privatisation may be known as the 'British Model' reflecting their more recent origins.

Why is it so controversial?

An American water manager once remarked that 'the closer you get to the customer the greater your chance to screw things up.' That is why my remarks about Shannon Information in chapter one matter, why fear and loathing are easily evoked by anti-private sector lobbies and why the greater the degree of private sector involvement the more the scope there is for fear and loathing.

During November 2011 there were two referendums regarding municipalities outsourcing their water treatment plants in North America.[8] In Abbotsford, British Columbia and Lodi, California, voters rejected proposals to involve the private sector, despite the Abbotsford project proposal costing C\$291 million against C\$328 million under municipal procurement and the Lodi project now costing the city a further \$900,000 each year. This gives an idea of the power of paranoia when you can oppose something on ideological grounds rather than on cost or service delivery. In Canada, there are few private municipal projects and so few examples for people to compare. The great irony here being that Canada's public sector pension funds are amongst the world's leading investors in private sector water companies from Chile to England.

In essence, water is unique in that a considerable NGO lobby believes that all water ought to be free irrespective of its actual cost, that access to free water is a human right. As we have seen before, they see it is as a 'Gift of God' and not a commodity and can only be owned on behalf of the people. There is also opposition to the idea of its being managed by foreign companies, along with the transfer of its management beyond current spheres of social and political influence and patronage.

Karen Bakker notes 56 'selected examples' of public protest against PSP from 1985 to 2007 (although I cannot remember those on the streets of England and Wales) pointing to their attraction to a 'green-red' alliance seeking to unite labour and environmental protest movements.[9]

Fear and loathing stalks the sector. There is a mildly amusing site called Powerbase that claims to be a 'guide to networks of power, lobbying and deceptive PR within the water industry'. My entry appears typical, being a rehash of old information and I suspect that the only people to be put out by this project will be those not listed.[10]

A brief history of PSP

As we saw in chapter three, the private sector and water management go back a long way. Beyond London, William Yarnold founded the Waterworks for Newcastle & Gateshead in 1697 and the first water distribution franchise in Paris was awarded in 1782, while the York Water Company has provided 'that good York water' to

York County, Pennsylvania, since 1816. A plethora of companies and contracts emerged in France, Germany, Italy, England and Wales and Spain during the second half of the nineteenth century and Lyonnaise des Eaux (Suez Environnement) and Générale des Eaux (Veolia Environnement) gained a number of international contracts between 1880 and 1918, after which these returned to state control. PSP had a low profile during the inter-war years until the leading French companies developed a presence in the former French West African colonies since the 1960s. Their international expansion has been related to French Government support. Aguas de Barcelona and FCC of Spain's international activities were originally based in Spanish-speaking Latin America. Thames (along with some of the other privatised water utilities in England and Wales) gained a series of contracts from the early 1990s, and other Spanish, Italian and German companies looked to turn contacts into contracts. In each case, the companies sought to diversify from their domestic contract base by exporting their management experience in the sector.

A statistical interlude – PSP in numbers

Since 1999, I have written the *Pinsent Masons Water Yearbook* and have been tracking how the private sector has emerged. Back then, I identified 73 private sector companies active in this market, 63 from developed economies and ten from the rest of the world. By 2011, this list had grown to 167 companies, with 84 of these in developing economies. A further 121 small and local companies have been identified from 22 countries, 111 of which are in the developing world. I have logged 1,209 privatisations and PSPs awarded between 1987 and the end of 2010, serving some 786 million people.

The private sector never quite faded away before its recent revival. Indeed there are asset owning companies in the USA serving 25 million people and 14 million people in England and Wales along with concessions and O+M contracts serving 71 million people in France, Spain, Italy and Germany along with 36 million served by O+M outsourcing contracts in the USA. With 155 million people served in the 'traditional' markets and 715 million by contract awards and privatisations (net of contract losses) and perhaps 25 million more through small company contracts and further service extension and population growth, some 895 million people were served by the private sector in 2010 or 13% of the global population.

Despite some excitable headlines, private contracts may end for quite prosaic reasons. From 1987 to 2010 contracts covering 15.4 million people ended unilaterally, either because the company pulled out or because the municipality ended it. Thirty-five contracts involving 46.7 million people ended by negotiation, which can range from a face-saving way of being booted out to selling the contract back to the municipality at a worthwhile price. Finally 21 contracts covering 8.5 million people have ended having reached their expiry date. Contracts can be re-awarded after their expiry and this has also happened with some contracts that ended by negotiation. While contracts ending unilaterally lasted 5.0 years on average, those ending by negotiation last 6.4 years and those at their expiry date, 7.7 years. The latter is affected by the number of short- to medium-term management contracts included. In total, I estimate that contracts covering 71.4 million people had been ended by 2010 or 7.5% of all contract awards by people served.

A decade ago, there was something behind the assertion that the market was being 'gobbled up' by major international companies. Indeed, in 2001 73% of those served by private sector contracts were served by one of the leading five companies, all based in France, Spain and the UK. Today, those companies account for just 31% of the market. A plethora of local players have emerged especially in developing economies such as Brazil, China, Russia and India, which have become the real force behind the private sector's expansion in recent years, accounting for over 60% of contract award volume since 2000.

Two ways of looking at water management

What do public and private utilities look like? Karen Bakker developed two paradigms 'Municipal Hydraulic' and 'Market Environmentalist' as a way of understanding urban water services.[11] My observations overleaf are meant to starkly illustrate what is wrong and what is right about the municipal and private sector approaches. Both approaches are currently evolving, sometimes for the better.

The 'Municipal Supply' utility is municipally owned and operated either serving people as citizens or not at all (non-persons) with services focussed towards those with political influence. Citizens may or may not have the right to elect those with ultimate authority over the utility although water provision is typically a low priority amongst politicians. As a natural monopoly, a water utility is a proxy for exercising power and political influence at various levels of society through job and business awards. It follows that utilities primarily exist to provide employment and political influence rather than a service. The appointment of its management can also be subject to political allegiance. Water management is a matter of making sure that stakeholders who matter get the water they want. It is supply led and designed to manage the status quo unless otherwise directed. Bakker noted that under the 'Municipal Hydraulic' paradigm, water losses rose as they were neglected, cancelling out the new infrastructure while sewage treatment works tended to remain on the drawing board, increasing the environmental impact. In the worst cases, corruption can be endemic as they offer opportunities for corruption in service provision such as obtaining a connections or avoiding billing and tariff collection. There is also the corrupt procurement of goods and construction services added by a lack of transparency through cross-subsidies and minimised reporting to maximise headroom for financial manipulation. As there is no political support to invest in water or to charge appropriate tariffs, there is no incentive to maintain services beyond a level tolerated by those who exercise authority, let alone to extend or improve them.

Yet municipal utilities also have a duty to serve the public and civic society and a duty to promote the economic and environmental well-being of the area served. The electorate (and society in general) has the potential to force change upon an under-performing utility, so they need to be responsive to public concern. They can have access to low-cost and long-term funding sources improving the affordability of long-term investment.

The 'Private Regulated' utility is either municipally owned or privately owned (privatised) and privately operated, serving customers. Supply management predominates due to companies being incentivised by higher returns on a higher asset base and volumetric sales rather than for managing demand. The utility's overriding aim is to optimise shareholder value, while the regulatory process dominates all other stakeholder obligations. This motivates companies to exploit a natural monopoly through unnecessary tariff increases, to provide what is needed to keep regulators happy but no more, to run down the infrastructure to maximise short-term returns, to develop assets with future returns on them in mind rather than for their need, and to minimise the disclosure of information about their activities to optimise returns. By bidding low to gain a contract (dive bidding) there is the opportunity to boost returns in post-award renegotiations. Contracts may be gained through corrupt bidding and there are also opportunities for corruption in systems and services procurement.

However, private sector water companies have lasted for centuries and are aware that seeking short-term shareholder value can destroy longer-term value. A company has a duty to develop the contract to maintain shareholder value, which means they have to satisfy stakeholder needs and desires beyond the regulatory minimum in order to retain the contract and to gain new contracts. No contract is an island and if a company wants to gain other contracts it has to have a good reputation. Companies are apolitical; people are simply potential customers and the more customers they can serve, the better – which drives service extension. Tariffs have to be affordable since there is no reward in levying tariffs which don't get paid. The private sector offers efficiency and innovation for its public partners that can drive down costs, allowing funding flows to go further.

Then there are the NGO lobbies. The primary purpose of an anti-private sector NGO is to exist by having an entity or ideology to campaign against. To this end, they exist to generate funds for their own well-being and to promote their interests above all others irrespective of their merits. These NGOs can also ensure that services are subject to appropriate external scrutiny and to be a voice for those who are otherwise unheard. The role played by NGOs that are water charities (as discussed in chapter seven) is admirable and quite separate from these lobby groups.

What actually happens?

1997 was a vintage year for major concession awards around the world and the fate of three of these, in Manila, Casablanca and La Paz and El Alto deserve a closer look. Many of them used funds and advice from the Global Partnership on Output-Based Aid (GPOBA) programme, which seeks to make subsidies for extending basic services to the poor in developing countries work in an effective manner.

Cross subsidies in Casablanca

Lyonnaise des Eaux de Casablanca (LYDEC) led by Suez (France) was awarded the 30-year Urban Community of Casablanca (UCC) concession contract. Cross subsidies were used from the outset to extend the city's services, so that during 1998–2009, water and sewerage accounted for 27–35% of LYDEC's turnover and 60–70% of investment, reflecting the need to upgrade and extend the city's water and sewerage services. People pay on a rising block tariff, with the first 8 M^3 per month costing 2.92 Mdh/M^3 and extra use rising up to 13.25 Mdh/M^3 for more than 40 M^3 per month. Most of LYDEC's water is bought from ONEP, the National Drinking Water Administration for 3.95 Mdh/M^3 meaning that water for essential use is directly subsidised by LYDEC. As a result, 50% of customers pay less than $3 per month. By 2004, water leakage of 25 million M^3 pa had been repaired, equivalent to the needs of 800,000 people, with the number of connections rising from 440,000 to 710,000.[12] In 2007, a plan to connect 85,000 houses in slums started. Before 2005, it had been government policy to eradicate slums ('non-authorised' housing) and move people to new settlements.[13] With the appreciation that these people have rights, it was decided to mix upgrading housing and connect it to utilities with appropriate relocation for a further 50,000 households.

LYDEC sought to connect 85,000 households to water or sewerage by 2012 for $167 million, with newly connected households paying $3.75 each month over seven years, a total of $316, or 19% of the project cost. The $70 million project deficit is being met by the concession company and aid agencies. In one pilot project GPOBA provided a subsidy working out at $563 per new connection ($162 for water and $401 for sewerage) or 28% of the total connection cost for 6,128 households.[14]

Manila Water – serving formal and informal communities in Manila

Manila Water was awarded the concession for Eastern Manila in 1997. 40% of those within the Metro Manila area currently live below the poverty line.[15] The Tubig Para Sa Barrangay (TPSB) programme is designed to extend services into the barrangays (informal settlements) in the zone. It allows families to share the cost of a single water meter. Vended water typically costs 100 pesos per M^3, seven times higher that charged by Manila Water under the scheme. By 2010, this scheme was providing water to 1.7 million people, none of which had tap water in 1997. At the start of the programme in 1998, there were 39,000 cases of diarrhoea in the service area, compared with 22,000 in 2005. During 2004, a 36% reduction in infant related mortality due to diarrhoea occurred due to the improved availability of potable water in the zone. Since 2007, all water supplied by Metro Manila has satisfied potable water standards along with universal 24-hour water availability, against 26% availability in 1997. Households connected rose from 325,000 in 1997 to 1,193,000 by September 2011, or four million more people being served. Service extension has been attained through improved efficiency, billing and cross subsidies. Over the first fourteen years, non-revenue water has fallen from 63% to 11% and the number of staff per 1,000 connections from 9.8 to 1.4, with commercial and industrial water providing 46% of revenues while using 25% of the water. Foreign currency exposure has been kept down, with 49% of the company's loans in pesos.

The 2007 connection fee was Ps 7,531 ($163), which was split into 24 monthly instalments of Ps 313 ($6) on top of the Ps 200 ($4) per month average water bill. This worked out at Ps 513 ($10) out of Ps 8,513 ($182) in monthly expenses. With water costing Ps 12 per M^3 against Ps 175 per M^3 for water from a drum, the connection fee is paid for after 45 M^3 of vended water purchases has been avoided. For poor households, the average consumption is in fact 5 M^3 per month. GPOBA subsidies have been used in a series of projects for those who cannot afford these connection charges, so that households pay for the meter and guarantee deposit (Ps 1,620) and a subsidy covered the connection fee (Ps 5,911), with the deposit fee repayable in 36 monthly instalments. The GPOBA project connected 10,642 households in 2008.[16]

Bolivia: Low cost approaches in La Paz, and political priorities

A 30-year water and sewerage concession for La Paz & El Alto serving 1.48 million people was awarded to Aguas de Illimani (AISA) led by Suez of France. While 45% of the population of La Paz live below the poverty line, the proportion in El Alto is 73%. By using labour provided by customers, the cost of connecting poor areas was reduced from the Government standard connection fee of $455 to less than $315. Connection costs were repaid by the community over five years via an interest free loan, allied with micro credit for internal plumbing with projects only going ahead with at least 60% community support. Families not connected to the network pay an average of $4.78 per month for water against an average of $1.55 per month for connected families. By the end of 2005 a total of 97,031 families (608,000 people) had been connected to the water network.[17] There were disputes about the speed of the new connections and how to respond to an influx of new urban dwellers near the concession zone and, in 2006, Bolivia's President Morales created a water ministry charged with renationalising water operations, lead by Abel Mamani, who previously ran Fejuve, the anti-private sector pressure group operating in La Paz and El Alto. Sisab, the Bolivian services regulator, was called to produce an audit justifying ending the concession, but they gave AISA an A+ rating in April 2006 and qualified it as Bolivia's 'best firm'. This was overruled and the concession was handed over in 2007.

Iconic events...

The case of La Paz and El Alto is widely heralded by NGO groups as a victory for the people against a foreign 'grab' of their water. There is considerable literature purportedly reporting tales of corporate wrongdoing in the water sector. A depressingly high proportion of this material bears little resemblance to what in fact happened, which is a pity since there are plenty of instructive examples of private sector shortcomings. Quite a few of the contracts that ended early simply deserved to fail because they should never have been awarded in the way they were – the private sector, like the public sector, can shoot itself in the foot with great aplomb. Here are some famous examples.

Mistakes, as tragedy...

The Bolivian city of Cochabamba deserves its iconic status, because it encapsulates how not to develop and implement a concession. The fast-growing city has suffered from water shortages for decades. In 1997, the World Bank supported a concession serving the city but its lower cost option was rejected because influential landowners did not support it. The World Bank withdrew and in 1999 a 40-year $270 million bulk water, water and sewerage concession was awarded to International Water's Aguas del Tunari, the only bidder. The concession was then developed through negotiation offering a 15–17% guaranteed US dollar rate of return, with exchange risk covered by tariffs along with a law obliging everybody to connect to the system – including well owners – irrespective of their ability to pay.[18] No public consultation was taken either over the law or the concession process.[19] The World Bank wanted an alternative PSP proposal with lower capital costs and not needing any tariff rises for five years[20] but Aguas del Tunari pressed ahead. Tariffs were raised by 20% (and more for the poor) in January 2000 with a general belief that further price increases would follow.[21] That April, International Water withdrew after some days of serious rioting. The Bolivian government acquired 80% of Aguas de Tunari from the consortium for 25 cents in 2006.

Impossible conditions, such as unrealistic demands being placed upon current and future customers (even when the bidder does not appreciate these at the time) hardly encourage further investment and can be particularly important when seeking to make a concession politically acceptable.

...and farce

Enron today is a byword for corporate criminality; its Azurix water subsidiary ought to be a byword for corporate aqua-asininity. Azurix offers a plethora of reasons not to plunge into the water sector without looking from absurdly high winning bids for contracts to wonderfully inappropriate management strategies. I remember talking to the Investor Relations executive at Azurix in 2000 as she floundered over their commitment to spend $10 billion on water projects. Azurix had just set up 'water2water.com', which would transform the business of water rights trading worldwide; it was a bubble world.

Azurix spent $439 million in 1999 on a concession serving two of the three regions of the Province of Buenos Aires. The contract got off to a shaky start when the British management arrived on the anniversary of Argentina losing the Falkland War and the local media referred to this as the 'British Invasion'. That was its high point. During 2000 there were problems with contaminated water (and contaminated language), which were eventually dealt with, but in February 2002, Azurix cancelled its contract with the Buenos Aires province and rapidly sold off its other contracts. Water2water.com is still there today but not as part of any bubble world – it's a company selling accessories for ornamental fish ponds.[22]

Foreign currency debt versus local currency tariffs

I am somewhat bemused about how some companies, with a long track record in international business, have run into major difficulties with foreign-exchange risks since in the long term it is clear that foreign exchange crises are to be expected.

Suez and Aguas de Barcelona's gained the central Buenos Aires concession in April 1993 with a bid 27% below the previous municipal tariff. The Aguas Argentinas consortium inherited a network with 45% distribution losses, providing water to 70% and sewerage for 58% of the city. Aguas Argentinas was regulated by ETOSS which was staffed by former employees of the state company and not formally qualified for their new roles.[23] Since 1996, AA and ETOSS were involved in contract renegotiations over bill collection and charging and total investment for the first five years was $750 million against the $1,200 planned. From 2001, ETOSS imposed a series of fines relating to AA's performance as the company reduced spending in the wake of the 2001 economic crisis and the 2002 Peso devaluation. When the Peso collapsed, Suez had €480 million in hard currency debt there mostly in water projects. Over the next two years, the consortium made €969 million in provisions and write downs against the contract and foreign-exchange losses. Suez and Agbar sold their stakes in Aguas Argentinas to the municipality in 2006.

Similarly, Suez's Maynilad Water was awarded the western half of the Metro Manila (MWSS) water distribution concession in August 1997. Maynilad Water took on 90% ($800 million) of the city's foreign debt which as the peso collapsed doubled in local currency terms from PHP20 billion to PHP40 billion between 1997 and 2000. Suez sought to end the concession in 2003 and after two years of negotiations they sold 84% of the company back to the city, releasing Suez from $220 million of loans. Two years later, Maynilad Water was sold to two Filipino companies who have used local currency debt to refinance the contract.

So, currencies do indeed go up and down against each other, especially over a few decades. When funding a project with a hard currency that is funded by soft currency tariffs, surely that matters.

Contracts and economies change

For some decades, Malaysia has sought to develop a national urban sewerage network but by 1990, access to household sewerage had increased to just 5%. As a result, the Sewerage Services Act was passed in 1993 and a 30-year concession covering Malaysia's urban sewerage was awarded to Prime Utilities' Indah Water Konsortium later that year. A sewerage surcharge on customers' water bills was not popular and unlike the water component, payment was not compulsory. During the 1997 Asian economic crisis the government ordered a 30% cut in sewerage charges for commercial customers from the start of July 1998, driving the company into losses in 1999. The next year, Prime Utilities sold its Indah Water shares back to the Government for 43% of what they bought them for in 1993.

In Thailand, United Utilities gained a 41-year design and build and operate contract worth £160 million for a sewerage system serving 0.7 million people in Bangkok in 1993 to create the city's first wastewater collection and treatment system. In 1994, the municipality cut the number of working hours allowed each day, while calling for the system to be further extended, along with more interception chambers and manholes. By 1997, United Utilities had made £90 million in provisions against cost overruns before the Asian financial crisis later that year ended the project's funding with just 80% of the work completed. Legal arbitration over recovering these costs is understood to be continuing to this day.

Guaranteed returns are not guaranteed

Thames Water gained a contract in 1995 to build and manage a water treatment plant at Da Chang in Shanghai, providing drinking water for 2 million people. The contract offered a 12% guaranteed return on investment. In 2002, the State Council passed a law requiring all fixed return water contracts held by foreign entities to be restructured and Thames sold its stake back to the municipality at a profit. Guaranteed returns don't offer a reasonable balance between project risk and return, especially when customer revenues are used to ensure these returns, and it has turned out that in all but one of the cases where guaranteed returns were used for international companies the contract has ended prematurely.

And how are La Paz, El Alto and Cochabamba today?

You rarely hear about what happens after a private sector participation contract is terminated. In La Paz and El Alto, from 2007 and 2010, Epsas, AISA's state held successor connected 8,000 families to water and 6,667 to sanitation each year, 36% lower for water and 29% for sanitation than AISA managed.[24] Epsas's total of 295,000 water connections in 2010 equates to 92% water coverage along with 69% sewerage coverage of the post 2007 coverage area, in contrast to AISA's 99% water and 80% sewerage coverage in 2005 for the original area. To the activists, AISA's demise was a triumph. To the pragmatist, perhaps Epsas is finding AISA quite a hard act to follow.

Despite its status amongst the anti-privatisation movement, Cochabamba's water services have not enjoyed a renaissance under popular control. Given the backlash against tariff rises in 2000, financing has been a bane ever since and water truckers continue to dominate in the poorer areas, selling untreated water to a captive audience. In 2004 more than 85% of people in poor areas continued to lack access to piped water against 11% in the wealthier, more politically influential districts. The poor, as usual, end up paying up to 400% more for their water, with 64% of these people identifying their poor water resources with illnesses. By the end of 2006, 50% of all people in Cochabamba lacked piped water and of those with piped water most were limited to just three hours a day.[25]

Experience matters

For companies, the chief concern is to attain a good standing which will enable it to obtain similar contracts in the future, all of which will generate suitable profits. For the countries concerned, projects need to deliver their service delivery and extension objectives and to demonstrate that the country represents an attractive place to invest and operate in. During the 1990s, some companies sought to enter the sector without any prior experience. The most visible example was Enron's Azurix as discussed previously. Jim Winpenny took a look at various recent market entrants in 2006 and concluded that those without a background in the sector (either as service providers, engineers or constructors) tend not to fare well or endure.[26] It is not a market to muscle into; it is one that only makes sense to become involved in if you have something to offer, an expertise that your client does not enjoy. Indeed, one of the most recent trends is for outsourcing of a specific management of your operations, ranging from managing capital programmes (Glas Cymru in Wales) to operating a desalination plant for a utility

(PUB in Singapore) to improving the treatment and recovery of sewage sludge (plenty of examples in the USA).

Because most assets lie 'below ground' (pipes, sewers and so on) when a company successfully bids to manage these assets, what they bid for and what they have to deal with may be quite separate things. Getting a contract off the ground, so to speak, can be far slower than a company thought possible when bidding for it. There is no doubt that companies can be over-optimistic when making a sales pitch, but what matters is how they respond to the gritty reality they have to deal with.

Delivering what is needed

Quite a few studies highlight efficiencies attained by the private sector (eliminating excess staffing levels, competitive procurement of goods and services, using technologies designed to attain the required service outputs for the lowest cost and incentive-based legislation that rewards operators) with the result that, for example, in England and Wales privatised infrastructure companies have reduced unit operating expenditure by some 1.25 to 3.5% per annum more than might have been expected in the absence of a 'privatisation effect' which was still occurring twelve years after privatisation.[27] Looking at a number of O+M contracts in developing economies, access to and hours of water supply before and after the contract award shows how matters can be improved over a few years.[28]

Pre and post PSP performance in some water utilities

Contract (Operator)	Households with piped water		Hours supply / day		Period
	Before	After	Before	After	2000–05
Amman, Jordan (Suez)	90%	100%	4	9	1990–05
Barranquilla, Colombia (Tecvasa)	60%	89%	19	23	1995–05
Cartagena, Colombia (Agbar)	74%	95%	17	24	1995–05
Senegal (SAUR)	59%	73%	16	22	1995–05
Zambia (SAUR)	100%	100%	13	18	2000–04

Perfection may not have always been attained, but were there protests and campaigns against these improved levels of service delivery?

How does PSP affect public health? In Argentina child mortality fell by 5–7% during the 1990s in areas of Argentina where water services were run by the private sector compared with those where services were still run by the public sector, with a 24% fall in the poorest municipalities.[29] This can only work when affordability is factored into such contracts.

Despite some concessions not being able to meet their contractual targets, in 36 major water contracts surveyed in 2009, 24.7 million people benefitted from new connections.[30] In some cases this was due to the linkage to grant funding, most notably in Senegal or the use of tax revenues as in Guayaquil (Ecuador). Indeed, failings in service extension are often linked to a dependence on internal cash generation, which may not be sufficient (as seen in Argentina and Bolivia) to meet the anticipated targets. Improved efficiency allowed more direct investment to be funded directly through internal cash generation as in Côte d'Ivoire and Gabon, while better utility management encouraged increased donor funding in Senegal. In the latter case, PSP and good management has allowed the utility to directly attract investment grade sub-sovereign funding.

The World Bank looked at 141 utility PSP contracts and compared them with 836 state owned entities over at least five years between 1992 and 2005.[31] They found that in water services a 22% fall in the number of staff post-private sector involvement was matched with a 12% increase in connections or a 54% rise in connections per employee and that distribution losses were 23% lower. In sanitation contracts, there was a 37% increase in connections per employee and a 19% increase in residential coverage. All these were significantly higher than in the comparable state entities. The greater the degree of private sector involvement, the greater the improvement in productivity and service quality. Since every utility has unique circumstances, there is a limit to how far such comparisons can go.

Putting things into a broader context, studies which compare PSP with corporatised municipal utilities may be missing the point, since the latter are operating as quasi-PSP entities.

Capacity building can speak volumes

Implementing capacity building, effective regulation and tariff reform well before private sector involvement, the process has a much greater chance of working. Expecting people to pay more without improved services or service extension will rarely be popular. In Chile, regulations allowing full cost recovery for water utilities and setting the state's responsibilities in overseeing a contract started in 1988 so that capacity building developed a decade before the PSP process formally began in 1998, these conditions had broadly been satisfied. More than 90% of the population were already connected to water and sewerage services and sewage treatment rose from 17% in 1998 to 83% in 2008, with $3.5 billion being invested in infrastructure between 1999 and 2009. The entire process raised $2.3 billion, with the state retaining 35% of the sector's equity. In Chile, government intervention ameliorated problems when, after demonstrations in April 2001, the government started a 'water stamps' scheme to allow low-income residents to recover part of their water fees.[32] The chief concern had been affordability and, in response, the state used the $80 million pa in dividends it received to provide a $54 million pa subsidy for the poorest 17% of the population.[33]

There are several examples of major concessions helping a country to develop its water management capabilities with the aim of becoming a regional centre of excellence. In an age of centralisation, the creation of new regional entities matters, as seen in England and Wales since 1989. In the Philippines, Manila Water having been listed on its local bourse is now active in Vietnam, India and other parts of the Philippines. In Morocco, LYDEC is listed on the local bourse and the state owned ONEP is seeking to develop the country's experience with PSP[34] so that Morocco can, in turn, become a regional leader in technical support projects in Western Africa. When a company is listed on its local exchange, this transfers at least part of the shareholding in the contract from international to local hands and devolves corporate activities towards the local level. Locally listed companies can also act as a material boost to local bourses because a well-managed water utility is typically regarded as an attractive, low-risk investment. This is important in encouraging the wealthy to invest in their own countries rather than abroad and for attracting expatriate investors.

De-politicising slums

For the private sector, the political importance of people living in informal settlements is not the point; they are enfranchised as consumers. When people are willing to pay a fair price for suitable water supplies within a suitable regulatory framework, there is scope to develop a business case for doing so. When the regulatory environment makes it worthwhile for the concession company to invest in low-cost approaches towards serving informal and less well-off communities it can be a powerful tool for service extension. The company needs these incentives, as returns are usually lower than for providing water to more well off customers. This means service extension is easier where there is some wealth such as Manila Water serving the city's business districts and informal settlements. One worry is that this may become the 'cherry picking' of the most attractive contracts.[35]

While anti-PSP campaigners regard Jakarta as a 'failed' PSP, what do those who live there feel? People who live in 'kampongs', squatter settlements, traditionally depended on water vendors or water tankers. Between 1999 and 2000 Thames Water Pam Jaya connected the 12,000 people living in Marunda kampong to the mains supply.[36] Instead of paying Rp 33,300 for a cubic metre of vended water, they paid Rp 995–1,275, the cost of their water being subsidised via the overall concession. Likewise, between 1998 and 2009 in West Jakarta, under Suez the number of tariff payers (household and similar connections) more than doubled from 201,667 to 412,456. In the lowest three tariffs categories (the poorest households), tariff payers nearly tripled from 75,606 to 212,646 and there has been a significant levelling of service success across income brackets.

Why do the anti-PSP NGOs regard these contracts as failures? What is wrong about connecting the unconnected and extending affordable water provision?

Challenges to be faced

It is fair to say that PSP works best when everybody is happy with a contract. Getting to that point can be quite a challenge as these examples show.

Since the 1990s, it is clear that there is a real need for internationally recognised legal contract definitions to minimise the cost of developing contracts and the bidding process, which for a major concession can be up to $3–5 million per bidder.[37] In developed economies, a fully established set of concession rules and definitions, established by precedent lowers bidding costs. Here, established rules along with higher per capita revenues also make smaller contracts more attractive.

'Dive bidding' was mentioned earlier as an example of poor PSP practice. It describes a winning bid that has been priced simply to gain the bid and then to renegotiate the contract with the aim of forcing the regulator's hand.[38] Evidence of 'dive bidding' is usually anecdotal, but it is corrosive. Looking at infrastructure concessions that had been renegotiated in Latin America between 1988 and 2000, it transpired that 50% of the transport renegotiations were led by the concession companies, against just 6% were in the case of water and sewerage.[39] Even so, 45% of transport concessions had been renegotiated, against 71% of water and sanitation concessions.

There are some concerns about rising block tariffs being distorted as everybody benefits at the lower end so that the upper end needs to compensate for this.[40] The better-off may have private wells, while group purchases by less well-off people will mean their buying water at a higher price. More needs to be done in fine-tuning tariff mechanisms for both public and private service providers.

Why conflict?

There is a considerable body of anti-PSP literature, but with some exceptions it can be characterised by a lack of engagement with the practicalities of water and wastewater management and service provision from either a public or private perspective. This stems from a postmodernist world outlook where all truths acquire equal validity irrespective of empirical reality. As a result, there is an existential dichotomy between various NGO lobbies and those involved in the practicalities of service provision. This is an example of agnotological thought, agnotology being the 'study of culturally constructed ignorance' where 'truth can be the antidote, but it must be fought for, the truth often encounters powerful resistance'.[41] Other classic agnotologies include climate change denial, Princess Diana conspiracies and 'Obama birthism'. There are echoes in the decline of consensus, and bi-partisan viewpoints are being seen as tainted in American politics. If you subscribe to one ideology, you are expected to adhere to a collection of ancillary viewpoints, irrespective of their underlying logic. Looking at the anti-PSP and Full Cost Recovery media, this is a fair comment. There is simply no interest in engaging in a fact-based debate, because it is an ideological conflict, a battle of beliefs, rather than one that relates to realities. Scientific academe from engineers to biochemists is by nature pragmatic. It seeks to develop new ways of addressing old problems, irrespective of who uses them. Indeed, Professor Richard Bowen's 'moral engineer' is charged with developing applications that serve the good of society.[42] The social science side of geography does appear to operate on the presumption of scepticism towards the private sector.

Maude Barlow is one of the best known campaigners against the public paying for water and the abolition of private sector participation. Her book titles tend towards the excitable (*The Fight to Stop the Corporate Theft of the World's Water* and *The Fight for Water as a Human Right*), while PSIRU, a trade-union sponsored group favours the prosaic ('*The failure of the private sector to invest in water services in developing countries*'), but at least they both avoid the exclamation marks that are usually to be found in these publications.[43] Barlow's central proposition back in 2002 was that water should somehow be paid by industry, which would not go very far in many developing economies.

One of the difficulties facing water management is that to deal with our real and present-day problems (with plenty more in store for our children) is that we have to address a resource squeezed between free-market fundamentalists and anti-capitalists at a time when the broad thrust of the academic debate about water policy appears concerned with a series of theoretical dialectics. I am sure these people have much to add to debates, but they need to acquire a language mutually comprehensible with those involved in the practicalities the sector requires. The improved transparency that typically takes place when a private sector operator is involved means that shortfalls (whether new or pre-existing) are exposed to a scrutiny that did not hitherto exist.[44] Anti-PSP reports constantly distort information by deluging data sources with papers deliberately designed to put PSP in a poor light, irrespective of underlying realities.

When I was preparing a paper for UNCTAD's 2008 *World Investment Report*, the draft of the main report included an assertion from PSIRU that the private sector had extended services by one to one and a half million connections. Since, in reality, just 36 contracts had increased coverage by 24.7 million people, this highlights how hard data gets submerged by other narratives. In a few moments at the 'Great water debate' in Zurich in 2010 Maude Barlow[45] managed to crystallise oppositional relativism; 'it would only cost $30 billion to hook everybody up to clean water for sanitation and drinking water', while 'in the private sector, tariffs line the pockets of investors, but in the public system they go back towards infrastructure repair, source protection and ensuring water for all.' The first remark is an underestimate by a mere 94–96% on the narrowest criteria I can use, while the second would intrigue water and wastewater managers worldwide. Yet when questioned about cost-recovery realities in Uganda she replied: 'Nobody's saying the private sector cannot, and in this case did not, improve the situation.' The corporation remains under public ownership and control.

For critiques of PSP I prefer to look at the experiences of those already actively involved in providing services. WaterAid, the UK water charity gets funding from the privatised companies in England and Wales and their staff but does not pull its punches when it believes private sector projects are not working. In Tanzania they highlighted how World Bank financing was conditional on a water contract, which was eventually rescinded.[46] Karen Bakker's two critiques of PSP (*An Uncooperative Commodity* and *Privatizing Water*) are by some way the coolest appraisals of how and why the private sector can and does on occasion fail to deliver.[47] I recommend both to those seeking a dissonant voice, as while we may differ on interpretations I admire her evidence-based approach.

Accusing capitalism of being inherently hostile to environmental sustainability is both asinine and self-contradictory. In its most primitive forms, capitalism assumes unlimited resources and the belief that pollution is irrelevant. This attitude sadly prevails in certain companies and politicians in the USA. In recent decades

pollution has increasingly been recognised as what it is –
a waste of resources caused by inefficient processes. In
chapter eight we saw the negative effects of pollution on
a socialist economy in China. Having spent some months
travelling in western China in 1988 to 1990, I saw for
myself the open asbestos mines and anoxic rivers bubbling
with filth. In the former Soviet Union, a substantial
underground literature has emerged ranging from
discharging oil and mining wastes into rivers[48] to unsafe
drinking water.[49] As Edwin Dolan put it, 'a market is not
truly free unless it includes effective mechanisms to
ensure that everyone...respects the persons and property
of others.'[50]

PSP for water appears to have become a proxy for
people opposed to capitalism who have had many of their
traditional outlets constrained by the collapse of the
Soviet Empire since 1989. The problem about proxy wars
is that they often involve collateral damage. In this case,
while I doubt if any of the anti-PSP / cost recovery lobbies
actually mean to do harm by making the business of water
more risky through undermining contracts and opposing
cost recovery, they may well be making it more expensive
to do good works by encouraging financiers and
development agencies to concentrate their efforts on less
contentious sectors.

Am I biased? I follow the private sector's failures as
closely as their successes. Perhaps I am a critically
sympathetic. When it comes to risk management, there
are plenty of case studies to write about. As a scientist, I
tend to regard the verifiable in favour of the sensational.
This, I appreciate, is anathema to many, but as a child of
the Enlightenment I prefer reality to conspiracy theory.

Suez, Veolia and Thames Water made a number of contributions to the debate in the past but corporate publications have been thin on the ground in recent years.[51] Much work has been carried out by AquaFed (the International Federation of Private Operators) and the UN's Special Rapporteur Catarina de Albuquerque[52] especially in relation to the human right to water, along with work developed for the World Economic Forum at Davos but attempts at rational engagement are drowned by a roar of implacable hostility.[53] There is a real need for an institution ready to act as a clearing house for information and case studies about how PSP has in fact performed. At least then the case for and against PSP could be argued using the same data.

With friends like these

You would imagine the capital markets would be cheerleaders for water finance, since there is so much business to be done. Unfortunately, the tensions between short-term and long-term outlooks and the modern need to 'churn' clients against a tradition of building relationships have made them a difficult place for water finance to function in. To give you a personal impression, let us look at the water sector in England and Wales between 1994 and 2000. In the summer of 1994, Ofwat, the economic regulator gave the companies what was seen as a generous price settlement for the 1995–2000 period. Shares fared well and analysts encouraged companies to capitalise on these good times by dividend hikes and share buy-backs rather than developing their businesses. This is what was called enhancing shareholder value. In 1995, Southern Water organised an analyst visit to Cowes Week. It was a long hot day, wining and dining across the country. I can't remember why I was talking to another analyst about the limits companies might consider. As he looked across from the yacht club's balcony, he sipped some fizz and observed, 'I love the purity of free markets.'

The Water Rats is a dining club for equity analysts following the water sector back in the 1990s. We still meet every Christmas, mainly as a get together to reminisce about the good old days when equity analysts were allowed to have opinions. Geraint Anderson the 'Cityboy' blogger is a member, and he has pretty scathing opinions about the City.[54] In 1997 I remarked to the analyst next to me how good it was that Welsh Water was continuing to support the Welsh language. I admit that was a silly thing to say in the City at the best of times, but he was rather tired and emotional, and spat back that 'the only function of a water company is to enhance shareholder value', followed by a stream of expletives, since 'nothing else matters'. I left the dinner early. The market force purists and shareholder value champions prevailed. Indeed, to this day, I believe it was the tone of their coverage that encouraged Ofwat to set an excessively harsh price settlement for 2000–05. Share prices fell to levels not seen for a decade, with Thames Water selling out to Germany's RWE and Hyder, the holding company for Welsh Water, imploding. According to one banker, a French water company executive told him, 'Ofwat is our best friend...they got rid of the British.'[55] This was how the 'teenage scribblers' of the shareholder-value movement managed to destroy the sector's value. Such an inability to relate hubris to nemesis was a telling pretext to the Internet bubble of 2000–02 and the current financial crisis. I tend to refer to reductionist shareholder-value analysts collectively as a 'thicket'.

Underinvestment means a harder edged capitalism

Where there is inadequate coverage, room is made for the unregulated cowboy capitalism of the vendors. Each advance in water management may result in job losses, from water bearers in London and water vendors today to sewage scavengers across history; pipes and sewers make these jobs obsolete, usually with great advances in human welfare. Utilities exist to provide human welfare; a service, rather than jobs. Throughout this book, the emphasis has been on what people need and why they need this, rather than what politicians and policymakers believe they need.

Higher standards and expectations also create opportunities. The International Finance Corporation has noted that where there is no access to tap water and / or safe sanitation, other approaches are needed until these gaps can be addressed.[56] They estimated that 704 million people worldwide use approaches ranging from boiling drinking water to filtering it, all of which creates opportunities for private companies when state utilities decline to provide comprehensive coverage.

Towards consensus

Celine Kauffman at the OECD points out that any project involving PSP needs to consider fully what degree of cost recovery is to be sought and what external sources of funding can be mobilised.[57] This is particularly important in sewerage / sewage treatment, as the benefits are not so immediately tangible to the consumer. Water conservation is a key element towards achieving sustainable cost recovery, since demand management (tariffs for encouraging sensible water use) as well as supply management (minimising distribution losses) can focus capital spending towards the most beneficial outcomes. She also points out that technologies can constrain management and, since circumstances and aspirations can change over the life of a contract, the more flexible the technology, the better. Finally, transparency in financial reporting needs to be encouraged, so that verifiable financial and performance data is easily available.

The World Bank's Philippe Marin concluded that PSP works best when it follows some simple rules. The private sector is not a 'magic bullet' but a tool for addressing specific challenges and developed with local circumstances in mind. This is why the recent shift towards more local PSP is both necessary and encouraging. Information is power, so the benchmarking of all utilities in a country works to the benefit of all. By planning for the long term, stable contracts that deliver for all can prosper. This can be done as companies get to know what assets they are operating in reality and develop a trust-based relationship with the regulator and show that they willing to be flexible. Efficiency works across the system. From gaining contracts through open bidding, using industry knowledge to buy the best value and most appropriate hardware and re-mobilising excess staff when possible, the company delivers value for all stakeholders. As the World Bank has noted, 'vested interests will naturally oppose any changes to the status quo that threaten their sources of revenue or political support.' So can you co-opt small private sector providers into a formal system by, for example, giving them opportunities to plug into the formal network in more marginal areas?[58]

Another way – Glas Cymru, Water and Wales

The evolution of Glas Cymru came as a response to extraordinary circumstances. Dŵr Cymru Welsh Water was renamed Hyder ('confidence' in Welsh) in 1996 after buying SWALEC (electricity and gas in South Wales, privatised in 1991). Hyder's multi-utility strategy was based on the assumption that savings and synergies would fund the SWALEC acquisition. That strategy fell flat when a 'windfall tax' was imposed on privatised utilities in 1998, and its tariffs were cut by 13% in 2000. DCWW was sold to the public for 240p per share in 1989, rising to 1,048p in January 1998 before falling down to 179p in March 2000. By that time, the cost of its dividends and debt meant that debts would exceed allowable levels by late 2001.

WPD, a US power utility, acquired Hyder in October 2000 and sold DCWW to Glas Cymru ('Green Wales') for

£1.85 billion, in effect Hyder's debts in May 2001.[59] Glas Cymru's management developed a bond-financed model to cut the cost of its financing. The re-financing lowered the net interest cost by £40–50 million per annum in the first two years of operation and ending dividend payments saved a further £58 million pa at 1999 levels.[60] While the company has invested heavily in new assets, operating spending rose by 1% in 2000–10 compared with 8–34% for the other companies.[61] As a not-for-profit company, Glas Cymru can distribute retained profits as customer rebates or extra spending to improve service quality or sustainability. Customer rebates have been used to ease the bills that were above the sector's average. From 2000 to 2015, their bills are falling by 6% in real terms compared with a fall of 3% to a rise of 17% in the rest of the sector. In a decade, rebates have cost Glas Cymru £150 million along with £140 million in discretionary spending.

What about their performance? Total leakage was brought down from 446 million litres per day in 1989 to 193 million litres by 2010, meaning that less water is needed to provide the same services, a useful example of demand management. When Dŵr Cymru Welsh Water was privatised in 1989, just 50% of sewage was treated to secondary standard and disposed to land, 30% was treated to primary standard and disposed to sea and 20% untreated and disposed to sea.[62] In 1988 10 out of the 37 designated bathing waters failed their tests, hardly a surprise with untreated effluents from a population equivalent of 690,000 being discharged straight into these waters. Between 2000 and 2005, new and upgraded wastewater treatment facilities serving a population equivalent of 2.56 million came into operation. By 2009, 99.8% of sewage was treated to at least secondary level and 29.6% to tertiary level.[63] In 2010, all 81 designated bathing areas passed the basic standard and 72 also met the guideline standard.[64]

I followed DCWW / Hyder as an equity analyst from 1989 to 2001 and was a 'Member' of Glas Cymru Cyf for the following decade, one of a group of unpaid external 'stakeholders' who liaise with the company about their performance. I believe the not-for-profit model has succeeded in driving down operating and financing costs and customer bills faster than traditional private sector utility management approaches, while ensuring that its customer service and environmental obligations have also been met. Glas Cymru ought not to be a one-off, but an alternative paradigm for utilities looking to optimise their operating efficiency.

Paradigms for water management

There is no room for complacency as the entire sector needs to be moved towards a more sustainable form of management, both in economic and environmental terms. The private sector has made much progress especially in environmental reporting and improving the overall transparency of the business of water management. Sustainability depends on depoliticising water management and concentrating on efficient service delivery irrespective of who manages the utility. There are three ways which can bring them about, none is better than the other and indeed, they will do best through cooperation rather than conflict.

Municipal Sustainable

Municipally owned and operated, although individual facilities or processes may be contracted out where a third party can prove that this is in the municipality's interests. All citizens are either served or will be served with appropriate intermediary measures already in place. Demand management lies at the heart of long-term sustainable water management, seeking to provide appropriate services at an affordable cost. The utility reports to a single government department as well as a municipality and operates on a sustainable cost recovery basis with full public reporting of the utility's services and finances. Singapore's Public Utility Board is a case in point.

Private Sustainable

Either municipally owned or privately owned (privatised) and privately operated, focussed on serving customers. The regulatory settlement is explicitly developed to encourage sustainability, accountability and demand management with incentives for innovation and efficiencies that enhance these aims. By its very nature, demand management should lie at the heart of this approach, since avoiding the need for extra assets drives costs down. Much of this is about grown-up relationships based on mutual trust, especially between operators and regulators, and about a balance between sustainability, economic efficiency and service quality that generates stakeholder confidence. The Companies such as Wessex

Water, Glas Cymru, Manila Water and Chile's Aguas Andinas show how this paradigm is developing.

Community-led management

Beneath the larger towns and cities, there is a serious case for Karen Bakker's proposed community-oriented management model, especially in developing economies, albeit bearing in mind her reservations about how far it can work in practice. I would build on it by saying a community can be flexible in choosing partners where these partners work in the community's interest and the community retains active control of the project and gaining access to funds, although issues of scale will inevitably arise. There is a fluid relationship between community management and private and municipal management, as communities are absorbed into other urban areas, they may wish to merge, or particular communities may prefer a distinct approach from other areas. Some of the small town projects in Colombia and the Philippines are examples of this.

It is desperately difficult to do these things well but they must be done well. The financial rewards for doing this are simply eclipsed by the humane rewards.

Conclusions – PSP has a positive role to play

Every politician ought to sacrifice to the graces and to join compliance with reason
Edmund Burke, *Reflections on the Revolution in France*

When I made a forecast of 21% worldwide PSP in 2025 at a conference, one delegate remarked that this meant '79% having crap service', even then. That is unfair, but so is much of the criticism of private sector involvement with water services. Today, people are dying due to the politicisation of water in a proxy war about outmoded political paradigms. Unlike money, politics and water do mix even when we would rather they stayed apart. If you conjoin the Chinese symbols for 'river' and 'dyke' you get the composite symbol for 'political order'.

Much of the difficulty in discussing the role of the private sector lies in flaccid clichés being uncritically accepted. For example, 'grabbing' water is a strange expression since water resources are usually in government hands, except when regarding local private supplies and owners of water rights. The utility has the right to abstract water as allocated by the government. If you try to grab water, you will find that it flows between your fingers. In a way, water is indeed valueless – until it flows from a tap, where it is needed, when it is needed and of an appropriate quality. Water without a price is water without a value.

What are the pros and cons regarding PSP and international water companies? Using PSP in this context allows for the mobilisation of new sources of funding and management experience and therefore assists governments struggling to mobilise new sources of funding. Against this is the inherent need for companies to make suitable returns on their investments and the need to balance the returns made against the benefits accrued by the government and the risks carried by both sides in undertaking such a contract. International companies have played a significant role both in the global impact of water PSP in developing economies and in assisting to deliver service extension. Can they be encouraged to make a still greater contribution towards universal access and sustainable management? While private operators have limited abilities to provide new sources of finance, they can greatly improve the financiability of a utility or its ability to attract new sources of funding.[65] The emergence of new local players has transformed the nature of the business, making something that had been in danger of being seen as a global cartel into a truly competitive business. It is also up to the private sector to engage more openly and actively in policy development, to address common misconceptions about water and wastewater and to support objective research into PSP's performance.

Debates about public and private services tend to overlook the role of regulation. Utilities need to be held to account, as part of the confidence-building process. To date, regulation has often been found wanting either because, as in England and Wales, regulation may appear to become an end in itself, seemingly generating data demands to justify its existence or because people have axes to grind such as the regulators for Aguas Argentinas, formed from staff sacked as the former municipality sought to professionalise its activities. Regulation and reporting matters, as it is the absence of incentives and oversight that holds back many municipal utilities, as seen in India, where there has been no need for adequate record keeping.

Unfortunately, there will always be an information arms race between the regulators and the utilities, as each tries to second-guess the other. The negative impact of this arms race can be mitigated by co-operative data disclosure with a more equitable and stable relationship as a reward. A regulator will naturally seek to be bigger, costlier, more powerful and extensive, and so regulators in turn need appropriate oversight and regulation. Independent regulation allied with appropriate resources and capacity building is essential to ensure the private sector operates to the interest of all. It has to ensure the optimal balance of service delivery and affordability while encouraging sustainable investment in the utility and its infrastructure.

Universal access will not happen without the private sector's active involvement. If the NGOs transferred their attention from attacking water and wastewater PSP for the sake of ideology to engaging with the private sector in order to make them perform more effectively, they could become a force for good. One thing and one thing only matters, that universal, affordable and sustainable access to water and sewerage access is attained and sooner rather than later. Opposing this surely is the prerogative of the morally bankrupt.

---

[1] JW, Personal communication, 2009
[2] Winpenny J (2003) Financing Water For All: Report of the World Panel on Financing Water Infrastructure, World Water Council
[3] Karen Bakker, personal communication, 2012
[4] JW, personal communication, 2006
[5] de Albuquerque C (2010) Human Rights Council A/HRC/15/31: Report of the independent expert on the issue of human rights obligations related to access to safe drinking water and sanitation, 29 June 2010: Page 16–17 Para 47
[6] UNCHR (2010) The Right to Water, Office of the United Nations High Commissioner for Human Rights, Geneva, Switzerland
[7] Ban Ki-Moon opening remarks at the Private Sector Forum on the Millennium Development Goals, New York, 22 Sept 2010,

http://www.un.org/apps/news/infocus/
sgspeeches/search_full.asp?statID=951
[8] GWI (2011) Educating the masses, GWI Bulletin, 24th November 2011
[9] Bakker K (2010) *Privatizing Water: Governance Failure and the World's Urban Water Crisis*, Cornell University Press, USA
[10] www.powerbase.info/index.php/David_lloyd_owen
[11] Bakker K (2010) *Privatizing Water: Governance Failure and the World's Urban Water Crisis*, Cornell University Press, USA
[12] De Cazalet, B (2004) The role of Private Sector Participation in developing the water sector in the Mediterranean Region: The example of Casablanca, FEMIP Expert Committee, Amsterdam 25–26 October 2004
Suez Environment (2006) Sustainable Development Report 2005, Suez Environnement, Paris, France
[13] Cluzeau C & Benchakroun Z (2009) Access to basic services in shantytowns Casablanca – Morocco. LYDEC, Nairobi, Kenya 26–27 October 2009
[14] De Beauchene X C (2010) Morocco: Output-Based Aid (OBA) Subsidies for Water & Sanitation Connections in Poor Peri-urban Areas, Presentation to International African Water Conference & Exhibition 16th March 2010
Cluzeau C & Benchakroun Z (2009) Access to basic services in shantytowns, Casablanca – Morocco. LYDEC, Nairobi, Kenya
Djerrari, F (2003) Best practice in urban water resource management: Contribution of LYDEC in Casablanca, World Bank Water Week, Washington DC, USA, 4–6th March 2003
De Cazalet, B (2004) The role of Private Sector Participation in developing the water sector in the Mediterranean Region: The example of Casablanca, FEMIP Expert Committee, Amsterdam 25–26 October 2004
Cluzeau C & Mathys A (2007) Morocco: OBA Subsidies to Water & Sanitation Connections in Poor Peri-urban Areas – The GPOBA pilot and the full scale INMAE project in the Greater Casablanca
Suez Environment (2006) Sustainable Development Report 2005
[15] Manila Water (2005) Sustainability Report 2004, Manila Water Co, Manila, Philippines
Manila Water (2011) Sustainability Report 2010, Manila Water Co, Manila, Philippines
Manila Water (2011) 3Q 2011 Investor Presentation, Manila Water Co, Manila, Philippines
[16] Menzies I & Suardi M (2009) Output-Based Aid in the Philippines: Improved Access to Water Services for Poor Households in Metro Manila OBA Approaches, 28, July 2009
Almendras R (2007) Manila Water, Presentation to the 3rd South East Asian Water Forum, ADB, Manila
Manila Water (2005) Sustainability Report 2004

Manila Water (2006) Sustainability Report 2005

Manila Water (2007) Sustainability Report 2006

Manila Water (2010) Sustainability Report 2009

Manila Water (2008) Annual Report 2007

[17] Business News Americas (2006) Fejuve prepares to take over Aisa, BN Americas 7th September 2006

Suez (2005) Aguas de Illimani's achievements, February 2005, Suez, Paris, France

Suez (2006) *Water for All,* Suez, Paris, France

[18] World Bank & PPIAF (2006) Approaches to Private Participation in Water Services: A Toolkit, Appendix A, World Bank, Washington DC

[19] Slattery K (2003) What Went Wrong? Lessons from Cochabamba, Manila, Buenos Aires, and Atlanta. Water & Wastewater Annual Privatization Report, The Reason Foundation, Los Angeles USA

[20] Reid M (2007) *Forgotten Continent: The Battle for Latin America's Soul*, Yale UP, New Haven, USA

[21] Dalton G (2001) 'Private sector finance for water sector infrastructure: what does Cochabamba tell us about using this instrument?' Water Issues Study Group, School of Oriental and African Studies (SOAS), University of London (Occasional Paper No 37, 2001).

[22] www.water2water.com, accessed 1st September 2011

[23] Zerah M H, Graham K & Brocklehurst C (2001) The Buenos Aires Concession: The Private Sector Serving the Poor, New Delhi, Water and Sanitation Program

[24] GWI (2010) David Lloyd Owen: Some illumination on Aguas del Illimani, GWI, November 2010

[25] Forrero J (2005) Bolivians Find That Post-Multinational Life Has its Problems, *International Herald Tribune*, 15th December 2005

[26] Winpenny J (2006) Opportunities and challenges arising from the increasing role of new private water operators in developing and emerging economies, OECD Global Forum on Sustainable Development, OECD, Paris

[27] Europe Economics (2003) Scope for Efficiency Improvement in the Water and Sewerage Industries, Europe Economics, London

[28] Riskong K, Hammond M E & Locussol (2007) Using management and lease-affermarge contracts for water supply: How effective are they in improving water supply?, Gridlines 12, The World Bank

[29] Galiani S, Gertler P & Schargrodsky E (2005) Water for Life: The Impact of the Privatization of Water Services on Child Mortality, *Journal of Political Economy*, Volume 113, 2005

[30] Marin (2009) *Public-Private Partnerships for Urban Water Utilities: A Review of Experiences in Developing Countries*, World Bank, Washington DC, USA

[31] Gassner, K Popov A & Pushak N (2009) Does Private Sector Participation Improve Performance in Electricity and Water Distribution?, Trends and policy options No 6, The World Bank, Washington DC, USA

[32] Castro J P (2006) Water Services in Latin America: Public or Private? (Discussion of Four Case Studies), MSc Thesis, Erasmus University, Rotterdam

[33] Errazuriz P P (2010) Complete turnaround of Chilean water companies, Presentation to the GWI conference, Paris, February 2010

[34] ONEP (2006) Presentation to OECD Global Forum on Sustainable Development Paris, France November 29–30, 2006

[35] Budds J & McGranahan G (2003) Privatization and the Provision of Urban Water and Sanitation in Africa, Asia and Latin America, Human Settlements Discussion Paper Series Water-1, IIED, London

[36] Surjadi C (2003) Drinking water concessions: A study for better understanding public-private partnerships and water provision in low-income settlements, WEDC, Loughborough University, UK
Anwar A (2003) Regulating service for the poor, Jakarta, Indonesia. Paper at the PPCPP Session, 3rd World Water Forum, Osaka, 19th March 2003

[37] Budds J & McGranahan G (2003) Privatization and the Provision of Urban Water and Sanitation in Africa, Asia and Latin America, Human Settlements Discussion Paper Series Water-1, IIED, London

[38] Budds J & McGranahan G (2003) Privatization and the Provision of Urban Water and Sanitation in Africa, Asia and Latin America, Human Settlements Discussion Paper Series Water-1, IIED, London

[39] Guasch, J L & Laffont, J-J & Straub, S (2005) Infrastructure concessions in Latin America: government-led renegotiations, Policy Research Working Paper Series 3749, The World Bank, Washington DC

[40] Aquafed (2007) Practitioners' Views on the Right to Water, Aquafed, Brussels, Belgium
HDR (2006) *Beyond Scarcity: Power, poverty and the global water crisis*, Human Development Report 2006, UNDP

[41] Proctor R N & Schiebinger L eds (2008) *Agnotology: The making and unmaking of ignorance*, Stanford University Press, Palo Alto, CA, USA

[42] Bowen W R (2008) *Engineering Ethics: Outline of an Aspirational Approach*, Springer

[43] Barlow M & Clarke T (2002) *Blue Gold: The Fight to Stop the Corporate Theft of the World's Water*, The New Press, NY, USA
Barlow M (2007) *Blue Covenant: The Fight for Water as a Human Right*, McClelland & Stewart, Toronto, Canada
Hall, D & Lobina, E (2006) Pipe dreams: The failure of the private sector to invest in water services in developing countries, World Development Movement / PSIRU, London, UK

[44] Aquafed (2010) Avoiding misconceptions on private water operators in relation to the Right to Water and Sanitation, Submission to the UN Human Rights Council, 26th March 2010, Aquafed, Brussels, Belgium

[45] GWI (2010) *The Great Water Debate. Transforming the world of water: Global Water Summit 2010*, Media Analytics, Oxford UK

[46] WaterAid Tanzania (2002) Water utility reform and private sector participation in Dar es Salaam, WaterAid and Tearfund, London

[47] Bakker K (2003) *An Uncooperative Commodity: Privatizing Water in England and Wales*, OUP, Oxford, UK
Bakker K (2010) *Privatizing Water: Governance Failure and the World's Urban Water Crisis*, Cornell University Press, USA

[48] Komarov Boris (1978) *The destruction of nature in the Soviet Union,* Translated 1980, Pluto Press, London UK

[49] Pride P R (1991) *Environmental management in the Soviet Union*, Cambridge Soviet paperbacks, CUP, Cambridge, UK

[50] Dolan E G (2011) *TANSTAAFL: A Libertarian Perspective on Environmental Policy*, Searching Finance, London, UK

[51] Suez (2006) *Water for All*, Suez Environnement, Paris
Thames Water (2003) *Planet Water* 2nd edition, Thames Water Plc, Reading UK
Veolia (2007) *The right to water: from concept to effective implementation*, Veolia Environnement, Paris

[52] For a summary of their publications see:
http://www.aquafed.org/documents.html
http://www.righttowater.info/catarina-de-albuquerque/

[53] Aquafed (2007) Practitioners' Views on the Right to Water, Aquafed, Brussels, Belgium
WEF (2008) World Economic Forum Water Initiative Realizing the Potential of Public-Private Partnership Projects in Water, Davos, Switzerland

[54] Anderson G (2009) *Cityboy: Beer and Loathing in the Square Mile*, Headline, London, UK

[55] Jenny H (2011) Comment made chairing the South East Asia Business Forum, Singapore International Water Week, Singapore, 6th July 2011

[56] IFC (2009) Safe Water for All: Harnessing the Private Sector to Reach the Underserved, IFC, Washington DC, USA

[57] Kauffmann C (2009) *Private Participation in Water Infrastructure: OECD Checklist for Public Action*, OECD, Paris, France

[58] WSP (2009) Guidance Notes on Services for the Urban Poor: A Practical Guide for Improving Water Supply and Sanitation Services, World Bank, Washington DC, USA

[59] MBIA (2000) Information Memorandum: Multicurrency programme for the issuance of up to £3,000,000,000 Guaranteed Asset-Backed Bonds

[60] DCC (2002) Dŵr Cymru Cyfyngedig: Directors' report and financial statements for the year ended 31 March 2002

[61] Ofwat (2010) Financial performance and expenditure of the water companies in England and Wales 2009–10: Supporting Information, Ofwat, Birmingham, UK

[62] Schroder (1989) The Water Share Offers, Pathfinder Prospectus, J Henry Schroder Wagg & Co, London UK

[63] DCWW (2010) Dŵr Cymru Welsh Water June Return, Dŵr Cymru Welsh Water, Brecon, Wales

[64] Glas (2011) Annual Report & Accounts, 2010-11, Glas Cymru Cyf., Brecon, Wales.

[65] Marin P (2009) Public-Private Partnerships for Urban Water Utilities, Trends and policy options No 8, World Bank / PPIAF, Washington DC, USA

Chapter 10

Towards a New Vision

Rhetoric and realities

Apart from meeting and greeting people, giving speeches and taking notes, water mega-conferences are about assessing where we are now and where we want to go, via joint declarations.[1] Between 1972 and 2010 there were eleven such declarations about the need for safe water and sewerage. These communiqués shine a light onto decades of nations in negotiation.[2] Expressions such as 'water security' only appear twice, while 'water scarcity' has slipped away since 2001 and 'sanitation' has eased in frequency. 'Water quality', 'poverty', 'health' and 'science & technology' are almost ever present. Only 'climate change' has moved with the times, from a vague concern to a real and present danger – along with a general appreciation about how our activities are making water management problems greater. These conferences are a pleasant bubble world to live inside for a few days, but with the 2009 World Water Forum costing an estimated $200 million to stage, they can be detached from the broader realities.[3]

Reality matters, as in the non-debate about climate change and causality. Climate change denial simply is a dangerous and self-indulgent distraction. The sector is too busy dealing with the effects of climate change than to indulge in this debate. It is here and it is now. As with the 'water must be free' NGOs, their ideological purity is polluting other people's lives.

The reality is that despite formal water and sewerage networks having been pioneered some 5,000 years ago, in many parts of the world access to these services remain a pipe dream, even where they existed in pre-history. Exhortations and conference declarations change nothing unless acted on by politicians, as universal access is wholly dependent on water and wastewater provision being taken seriously.

We need to appreciate the challenges ahead...

Many of the problems facing urban water management now and in the near future may appear to be insurmountable. This is not being pessimistic but realistic. Until political attitudes evolve and policy blockages can be breached, progress will remain poor and, indeed, in many cases matters are set to get worse. We have seen plenty of this, from the remorseless deterioration of America's water and wastewater infrastructure to the unreliable nature of connection data in Africa and South Asia, as the ambitions of universal connection are mocked by promises for tomorrow and political indifference today.

Too often inefficiency and corruption have been taken for granted in recent decades as part of an unsustainable supply-led municipal hydraulic paradigm. The potential for making water and wastewater services financially viable and affordable depends on grinding corruption out of all aspects of service provision and procurement, ensuring that billing is comprehensive, comprehensible and fairly enforced so that efficiency acquires a moral dimension. By this I see the utility as an entity charged with providing sustainable and affordable water and wastewater services for all. This will require a great change in many management attitudes, but many should welcome a move from a culture of keeping politicians happy and working in near ignorance into one where you seek to inform stakeholders and you are better informed in turn. Financing priorities need to shift towards making funding go further and to encourage more investment in systems and services.

...which calls for honesty about management and money

Finding finance is a special challenge. It always will be until stakeholders are allowed to contribute towards funding network access and the right to economically participate in urban society. Sustainable and even Full Cost Recovery is cheaper for the poor than having to pay the informal sector's uninhibited tariffs. The business of providing water and wastewater is inherently financeable when utilities are allowed to operate within a reasonable political and regulatory framework.

Motivating management will always be a challenge, as they often have been conditioned to respond to others rather than to suggest new courses. Much of this conservatism stems from the difficult relationship between a utility and government and its various service, environmental and economic regulators. This requires a more honest, transparent and pragmatic relationship between utilities, regulators, governments and the mechanisms of their governance. This will be greatly enhanced by a free and full flow of information from utilities as a confidence building exercise. People are more willing to pay where they trust their service providers.

Beyond water, money can be mobilised for all manner of things. According to Brown University, the cost of the US interventions in Iraq and Afghanistan since 2001 will reach $3.2 trillion by 2015 and may be as high as $4.0 trillion.[4] That money, as we have seen, would provide basic urban water and sewerage systems for all people living in urban lower-mid and lower-income areas worldwide. Which intervention would have created the greater goodwill? In reality, translating aid into service extension is more complex than such a simple comparison, but it does say something about our priorities. Indeed, financing water and sewerage is about reasoned choices. Politicians prefer to spend more on defence than water and wastewater, and people will happily buy bottled beverages while opposing cost recovery for their water services. Meanwhile, in developing countries, people are allowed to spend on mobile phones but not on water and sewerage access.

Connecting utilities, cities and citizens

For too long, the water managers have been hobbled by poor information. We have reached the point when we can no longer endure manifestly false connection figures from countries whose governments and municipalities are more concerned with ticking boxes than what their citizens want and get. Myths have to be confronted and rebutted. Now is the time to embrace the information age. This means the comprehensive disclosure of how each utility is performing in terms of finance, sustainability and service provision along with measuring and monitoring water and wastewater as it flows through rivers, networks and pipes. This would create a world where performance data can be openly shared by stakeholders, utilities and governments in a position of trust and insights and feedback can be used to improve service delivery and confidence in the utility to provide this.

Water scarcity, a problem of our own making

The ability of a utility to distribute potable water depends on the quality and security of its water resources, so these need to be safeguarded. Difficulties with urban water management tend to be self-inflicted; without excessive population growth, supply–demand imbalances would not exist. Urbanisation and the localised depletion of resources go back thousands of years. Only recently has it become a widespread concern, exacerbated by people moving to water-scarce areas from Arizona to the Arabian Gulf.

419

Life evolved and adapted to the water resources available and when people migrated into these areas, they adapted in turn. Sometimes they adapted their environment, as in the water-fed cities and sometimes they constrained their numbers or water use. The argument that people are in the wrong place is somewhat disingenuous. Water tends to be in the right place, since people do have a habit of settling where water is found and developing cities where it abounds. It becomes the wrong place when a city outgrows its supply. Life on earth evolved to thrive in the habitats it encountered. We cannot complain about water shortages and scarcity, since living in high densities in water-short areas is self-evidently unsustainable. If we wish to live in these areas, we have to adapt to the water resources available.

Looking at water scarcity from a global perspective, Mark Lynas' *The God Species* considered various boundaries within which life on our planet can reasonably be sustained.[5] He assumes a global limit for human consumption at 4,000 Km^3 per year, compared with current consumption of 2,600 Km^3 per year. This is a reasonable figure, assuming that 10.8 billion people use as much water each as we do today, against current projections of 9–12 billion people by 2100. It points out that we have a limited room for increased water consumption – no matter how the global population rises and how water usage patterns evolve. To paraphrase Amartya Sen, 'thirst is the characteristic of some people not having enough water to drink. It is not the characteristic of there being not enough water to drink.'[6]

All being well, the world's population will stabilise at some point towards the end of this century before easing to a more sustainable level in the longer term. How many more generations will have to tolerate the service disconnects we put up with today?

Targets tend to miss the point

Target meeting, like other box-ticking exercises, may be useful in obtaining snapshots, but 'targetism' can obstruct the real benefits that were sought in the first place. Partial coverage misses the point that only universal coverage allows health benefits for all. That happens when you eliminate the presence of untreated excreta from communities rather than halving it, as the few who remain unconnected will always affect the many. Then there is the paramount role sewage treatment and wastewater recovery will play when it comes to sustaining the water cycle – neither of which featured in the Millennium Development Goals. The MDGs are a well-intentioned intermediate measure, but no more. It is time to address the real challenge, that of universal access to water and sewerage and to sustain the integrity of the water cycle.

Forget water wars; think about water peace

I have avoided the subject of 'water wars' since there is precious little evidence for actual water wars rather than disputes or conflicts.[7] Water wars are rarely in anybody's interest, despite Mark Twain observing that 'whiskey's for drinking, water's for fighting over.' Properly valued, water can become a tool for peace.

421

The tension between expectations and needs is a reconcilable one. The immediate need is for safeguarding human life and dignity. Even before people are connected to basic services, much can be done by ensuring that people have basic point-of-use water treatment. This requires honesty about just how many people are in fact receiving potable water supplies. From there, four approaches can be considered: provision of point-of-use treatment for safe drinking water, followed by getting the basic services in, laying the basis for future water and wastewater treatment systems, and laying down new infrastructure in peri-urban areas which will be developed in the medium term.

Planning ahead brings long-term benefits

Sustainable sanitation has to go beyond latrines and communal access; household access must be the accepted norm for all urban society. Likewise, foul and storm sewerage needs to be handled in such a manner that the integrity of rivers flowing through urban areas can be secured and that the water, nutrients and energy embedded in our effluents can be beneficially reused. Planners should assume that water recycling (greywater and rainwater) is to be expected and that potable and non-potable water systems can make sense, for example, in apartments.

Close the loop

The water-nutrient-energy nexus is going to play a profound role in demolishing the patronising assumption that only wealthy nations can afford sewage treatment and recovery. Wastewater is not just pollution; it should be seen as a neglected resource. Properly managed, it abounds with energy, nutrients and is a major source of urban water. This is a battle between populist prurience and prosaic reality.

Water reuse is an expression that needs to be carefully used. The 32% of Egypt's water which is reused, for example, is quite different from Singapore's 35%, since the former is at best partially treated sewage which is discharged for agricultural use, while the latter is about providing treated water to its customers. Water and sanitation provision offers limited long-term benefits if untreated sewage is discharged downstream. The urban water cycle has to be engineered to replicate the functions of the natural water cycle in order to deliver sustainable water management. By addressing sources of pollution, rivers from Singapore to Surrey have been returned to life in recent decades instead of being used as sewage dumps.

Water networks need smart water

The law of unintended consequences can pour through water systems management. For example, cutting domestic water use causes problems if the reduced throughput cannot flush sewage away. Demand management needs to reflect the requirements of the entire system. This is achievable, for example, by using greywater so that genuine efficiencies are gained, along with the flow needed to flush wastes through the networks.

A few years ago, smart water meant modern meters. Now it involves extracting data from right across the water cycle, making this data meaningful, so that customers and managers can use it through software programmes and applications. Smart water is about how the assets we have can be more efficiently managed and how best to develop future assets, especially when it comes to cutting energy bills. It is about showing how much energy goes (and costs) in a power shower as well as the water used, altering customers to unexpected water use in a building, from burst pipes to a neglected tap. Distribution losses can be hunted down by linking pressure management with flow control in both water networks, while river and sewer flooding can be predicted or avoided through real time flow monitoring combined with rain data. Real time pollution monitoring allows regulators and companies to respond to incidents immediately rather than dealing with the aftermath. Smart water will allow us to understand what we can learn from integrated layers of monitoring data, how water can be managed at the household as well as municipal level and how we respond to this information. Perhaps the smartest thing about smart water could be helping us to understand that water might indeed be valuable and ought to be sustainably valued.

In an interconnected world, river basin sustainability is for the common good

Transferring water from one river basin to another is of dubious value. Where basins have an 'excess' of water, it means that in fact they are less over-exploited and can be seen as reserves for biodiversity and water system health, rather than resources for the profligate to tap. The EU Water Framework Directive has a great ambition to return Europe's water basins to a 'good ecological quality'. Indeed, as with transforming Europe's beaches from being regarded as sewage discharge and dispersal zones a generation ago into places fit for healthy recreation, the principle of seeking to regain and maintain the integrity of Europe's river basins has profound implications for longer-term sustainability, both economic and environmental. European legislation is contentious and costly, but these directives offer the possibility of security of our future supplies.

Integrated water management and provision allows a variety of water resources to be used, eliminating resource overlaps. As seen in Singapore, rainfall harvesting can take place at the utility as well as the household level.

Innovation may bring salvation

Water is a risk-adverse, conservative sector, due to various political, public and cultural issues. There is vast scope for innovation in water management. Innovation is the art of the implementable and its integration. It is also about understanding what a technology is capable of delivering and ensuring that this happens. Too often, incremental improvements are dissipated through poor systems implementation. Regulators and politicians can be divorced from the science of water management, let alone the engineering side, meaning that they fail to understand how innovation can be encouraged and best deployed. This means keeping an open mind about supply options. It is easy to forget that desalination for a coastal settlement can be cheaper than dams or inter-basin water transfers, even more so for wastewater recovery. Pick the approach that fits the location.[8]

In the late 1980s, many believed that environmental spending would always rise and that markets would forever expand. It took a recession for economic reality to hit home and today, technocentrism has given way to pragmatism. There is business to do, when companies offer goods and services that allow water utilities to do more, to do it better and to save money at the same time. That is what sustainable innovation is about.

Bottled water shines a light on attitudes to tap water

Bottled water is usually no better or worse than tap water; it does not have any discernible health benefits, but it can make some people feel good. It is a value-added distraction, which should not be banned but shows how susceptible we are to fear, loathing and ignorance. Bottled water paradoxically demonstrates that water has a value, since people do, in fact, pay for it. This also reminds us that this money would be better spent on water infrastructure.

Water has to have a value in order to be valued

I have been repeatedly told that 'water is God's gift to man', which makes me wonder if sewage is the Devil's. If water is free, it is untreated, with cholera or dysentery thrown into the bargain. If water is to be free, who funds the sewerage networks and sewage treatment? The arid arrogance of the 'water is free' movements shows an almost callous disregard towards how water needs to be managed, let alone extending access and improving services. An air of unreality pervades where the Humanities and water management meet, with humanity appearing to be a low priority. Reviewing academic literature on water management on the sociology side of geography, it can appear that Marxist dialectic matters more than the simple act of service extension.[9]

Everybody has access to unreliable and dirty water supplies. It is affordable access to clean water that matters. That is why this water carries a cost, a different one to non-access. When it comes to paying for water, we can't afford not to have water for all. People, when given the chance to enjoy safe water and sanitation, will willingly pay a reasonable price for this. Compared with other utilities and human whims, safe water and sanitation are eminently affordable.

In Manila, Phnom Penh, Sao Paulo and Kampala, universal access to affordable water is becoming the norm. This has been achieved through a variety of public and private sector approaches which all have one thing in common: they serve the customer, not the politicians, plutocrats or unions. Phnom Penh's utility reforms have been popular because they provide genuinely comprehensive coverage, cutting the water bills of the poor by 70–80%. The rich resent Full and Sustainable Cost Recovery since cheap utility services were a traditional perk of the empowered. Funds available can be boosted by diverting state money away from bankrolling the corrupt towards actual service extension and upgrading work. For the developing world, a multi-staged approach can engineer in affordability, deliverability and long-term adaptability and sustainability. While my vision is for universal household access, physical obstacles do remain. Where in slums is the space for universal household sanitation? Well-built and well-managed communal blocks offer at least an intermediate answer. High tech, compact wastewater treatment and recovery systems can bring these services into densely populated areas using a minimal amount of space and operating costs.[10] These also use Full or Sustainable Cost Recovery in a simple and direct way via 'sweat equity' (offering your labour for free in lieu of funding) and by affordable user-pay management.[11]

Delivering desirable outcomes

What can an enlightened government do? The OECD summarised its desired policy objectives as 'Three Cs' namely: Commitment – against corruption and a desire to deliver service extension; Consistency – in policy development and implementation, free from conflicts between competing regulators; and Capacities – build capacity for regulation and implementation so that the benefits of capital spending justify their costs and that utilities have to justify their performance or face financial penalties.[12]

While price and cost data were akin to state secrets a few years ago, thanks to efforts from the UN's International Network for Benchmarking Water Utilities (IB-NET), GWI's annual tariff survey and Ofwat's annual international performance surveys, comparative utility data is now emerging. Utilities can start to meaningfully benchmark their performance with comparable entities across the world, reducing the scope for corruption and mismanagement while highlighting good practice.

Corruption can be tackled, with transactions challenged at every level. Corruption occurs wherever room for malpractice exists, so reporting and accounting has to be tightened to squeeze this out of the system. One way is to develop 'windows of transparency' during all procurement and tendering processes in order to winnow the scope for substandard or overpriced work. The same applies for inefficiency ranging from excessive staffing to wasteful procurement. For all its controversies, Ofwat has focussed attention on the amount of slack that existed within the water and sewerage companies of England and Wales. Since 1989, four price reviews have revealed progressively more room for efficiency.

Water realities mean that hope can hurt

Sadly, there is no realistic prospect of international aid agencies or governments (via Official Development Assistance or ODA) providing truly meaningful support for water and sanitation extension let alone sustainable water management. Water lacks the glamour of mega-infrastructure projects such as power and transport, let alone the limpid pleasures in upgrading a nation's defences. There is also the tendency for ODA to be directed towards the wealthier developing countries (as they are better customers) rather than those who really need it. It might sound a little obvious, but pro-poor subsidies should also go to the poorest 10–20% of the population, rather than the richest – as seen in Tanzania, India and Nepal.

Ways of delivering aid can be misplaced when it is used without consideration of circumstances or consequences. Dambisa Moyo's *Dead Aid* proposes that Africa needs to wean itself from aid dependency.[13] I fear she goes too far in proposing a post-aid future of private finance and liberalised trade, but it is increasingly clear that urban-directed aid ought to be a catalyst rather than an answer. Of course all donations ought to be gratefully accepted and likewise thoughtfully spent on small hard to fund projects in the worst off countries, not prestige schemes in wealthier nations. It ought to be spent where it is most urgently needed, where it can empower people towards developing economically self-sufficient water and sanitation services. Even more so where systems can be operated and maintained without undue dependence on imported parts and expertise.

This applies across the world, as seen in Northern Ireland where the politics of 'free' water (it is paid by the rates, which are in reality paid by the rest of the UK) has constrained the utility from addressing its service obligations. In 2010–11, 56,000 households lost their water supplies when pipes burst after a cold patch and customers were left high and dry in every sense. Without fiscal responsibility, management responsibility is also diminished.

While most developing countries never had full water connections (in recent millennia anyway), that does not excuse the status quo; without full water connections, developing nations will find it much harder to develop.

Instead, they have overlapping levels and forms of water service and means of access – from wells and rainwater, taps and tanks and vendors and bottles. It is time for civic society to look beyond satisfying its islands of affluence, what Karen Bakker refers to as the 'elite archipelagos': the gated compound approach of private wells, POU filters and a variety of bottles. I suggest that instead of the gated mentality, these islands could become the outliers of service extension, where water and sewerage mains are put in first and then services can be extended to adjacent areas. As with sanitation, get the skeleton network in and start with a network of community taps to get confidence in water quality and availability and then incentivise for proper household connection. Political will can go a long way, and developing sustainable water management is a key to economic development. Singapore gained universal household sewerage in roughly two decades following independence and universal wastewater treatment in the decade after. From being a developing economy in 1965, it has left much of the West behind in both affluence and water management.

Water is not about public against private, it is about serving people

Water and sanitation policy can be hijacked by special interests, from NGOs using it as a proxy when fighting capitalism to the wealthy and influential keeping these services for themselves on the premise that the poor and politically unconnected cannot afford these services. The belief that urban water services must be publicly held when there is little such debate over telecoms, electricity and, indeed, food is a triumph of ideology over reality. There is an ethical question about the legality of making a profit from something essential for life, but this also applies for food as well as clothing, shelter and heat. Involving the private sector ought to be seen as part of the package and where it can really help, it ought to be considered.

Utilities exist to serve their customers, not special interest groups, whether these are staff, politicians, management or shareholders. Utilities should have incentives to provide universal and high quality services and constraints to assure affordability to all. The overriding need to provide an appropriate service makes debates about delivery models irrelevant – what matters is the most effective model for the circumstances. Indeed there is great scope for hybrid delivery models, designed to deliver a utility's needs: for example, a corporatised utility with specific outsourced functions, where these are more efficient and offer better value than the status quo.

Factor in the human factor

A recurring theme has been how the human factor gets excluded from policymaking and political decisions. Research on the potential benefits of household water access has suffered from assumptions that people primarily exist to work and that any time freed up by improved access to water and sanitation will be spent working rather than enjoying life. Likewise people across the world are capable of making informed choices when it comes to spending on mobile phones; they are not a lumpen-proletariat that according to campaigners cannot be allowed to spend on formal services and have to make do with overpriced water from vendors, if at all. This means that mechanistic considerations about human nature need to be overcome and good intentions translated into a reality that allows for ideals to be met.

Mobility and utility

What else can the mobile telecoms boom tell us about water finance? Mobile telecoms work because they decentralised, freeing both the network and the subscriber from the tyranny of wires and therefore lowering the capital intensity of service extension. Can decentralised water and wastewater systems offer greater flexibility and minimise the mains networks needed to serve them? Likewise, what about treating wastes at the point of discharge when they are at their most concentrated? This might be a way of preventing drugs entering the water cycle as well as capturing nutrients and energy.[14]

Much that needs to be done can be done. Today's technologies are doing a good job and the pipeline of innovation I have come across points towards an age of progressive innovations. Few massive breakthroughs perhaps, but improvements which if suitably harnessed can make the principle of more being done for less cost (money, carbon, water) a continually improvable reality. When paradigms shift, assets need to move as well.[15] If water supply and demand become disconnected, especially due to climatic fluctuations, we need greater flexibility in mobilising assets. Instead of a gold-plated array of hardware lying idle in anticipation of a future event, perhaps it would be better to look at systems of pipes, pumps and treatment units that can be rapidly deployed when and where they are needed. The chasm between what will be done and what ought to be done can be bridged by declaring what must be done instead and setting hard timetables and targets that cut out the wriggle room. By choosing a 2050 horizon, the OECD raised the game; can anybody reasonably argue that such a fundamental need cannot be addressed in close to four decades?

A question of attitudes

While water has been treated and urban sewerage has operated for millennia, it is only in the last two centuries that the link between water and disease transmission has been appreciated. For example, Louis Pasteur changed attitudes towards water and disease in France. High mortality amongst the lower orders was assumed to be due to their poor morals and habits. Once water's role in disease transmission was understood, urban living could be transformed through proper sanitation.

The sector boasts a proud history of managers and innovators. Even today, many Londoners enjoy water sourced from Hugh Myddelton's New River Water Company project that entered service in 1613. Likewise, Joseph Bazalgette's London sewer scheme, completed in 1874, was one of many such projects that moved the *British Medical Journal* to proclaim sewerage the greatest medical advance of the past 150 years. It is perhaps inevitable that London's proposed 'super sewer' has been bogged down in a morass of special interest lobbies. Getting these grand projects done is essential if we want to sustain the health of our cities and the rivers that flow through them. Using innovation and efficiency to make all water and wastewater systems operate better is crucial. Policy in theory and practice needs to be about the art of the attainable. Politicians and policymakers should at the very least avoid inhibiting innovation and efficiency and should look at examples such as Malta's integrated smart water and electricity grid to appreciate the benefits good management can confer.

A vision for 2050 – sustainable, attainable and affordable

An immortal Work: Since Man cannot more nearly Imitate the Deity, Than in bestowing health.
Robert Mylne, Pedestal dedicated to
Sir Hugh Myddleton, Amwell spring, 1800

We started with declarations, so here are some quotations. 'The major problems holding us back...to where we should be in 2050 appear to be largely political and managerial bottlenecks, not technology or finance',[16] while according to Professor Seetharam Kallidaikurichi, water 'has to move away from being a construction-oriented industry into a socially-oriented industry. Then there's going to be huge business opportunities for them.' Asit Biswas believes that 'the issue...is political interference in the management of water utilities...we do not have a water problem. The problem is one of governance, the problem is one of management...during the next 20 years...the world of water management will change much more than it has in the past 2,000 years whether we like it or not.'[17]

The essence of this vision lies in the following nine points. They may appear to be somewhat simplistic and even idealistic. In no way does that make them any less valid.

1. *Be honest and realistic about what needs to be done and to what level. Don't worry about upsetting national pride, let alone special interest groups – humanity needs to prevail over ego and ideology.*

2. *Work out what these programmes will cost to develop and to manage and maintain. Use international best practice to identify high-quality and good-value work rather than those adopted by crony companies and corrupt procurers.*

436

3. *Inform all stakeholders about the task ahead and how it needs to be funded. Secure public support for the work needed and willingness to pay appropriate tariffs.*

4. *Ascertain how much funding can be raised through affordable tariffs and how much from debt financing (Full Cost Recovery), and whether tax revenues and aid (Sustainable Cost Recovery) can be mobilised.*

5. *Develop projects with their financiability in mind from the outset so that the financing is as low cost and predictable as possible.*

6. *Seek to eliminate corruption, inefficiency and malpractice by companies and utilities as a matter of societal and national honour.*

7. *Simplify the relationship between government, utilities and their regulators. Utilities ought to report to one regulator and one ministry in order to avoid conflicts of priority and in-fighting. Ensure that regulators are apolitical and fully trained, and consider international alliances of regulators to share best practice and expert staff. Use the regulators as a cornerstone of stakeholder communication.*

8. *Incentivise utilities to move from supply-led to demand-led management and, wherever possible, to carry this out at the river basin level and to integrate both water and sewage management. The water cycle is too important for conflicts between competing utilities.*

9. *Always place efficiency, affordability and sustainability above special interest groups and any political interference. All that matters is a utility that truly provides water services for all.*

These observations are universal; they apply to grand plans in advanced economies as well as to reforming and extending basic services in developing economies.

By 2050, I very much hope that population growth and urbanisation will be easing off in most countries. As the greatest demographic shift in human history eases into a new reality, ensuring universal access to water and sewerage services will be regarded by our children and grandchildren as an enduringly humane legacy. Utilities therefore need to reconcile what data is available with their need to develop and implement 20 to 50-year management plans. It is fair to say that those utilities with the greatest need to develop such plans are those that are currently in the worst position to do so.[18] Helping these utilities to develop the capacity to redress this has to be our overarching priority.

Obligations are interlinked. Governments need to collect the taxes required to support Sustainable Cost Recovery and to create regulatory and policy frameworks that allow innovative and efficient utilities to thrive. Utilities have to respect their obligations to deliver universal services, affordable and reliable water and sewerage for all. People have to demonstrate their willingness to pay for these services, to act as stakeholders and highlight poor service delivery and corruption. Companies need to develop goods and services that can further improve the efficiency and sustainability of water services. Like the water cycle, all these elements depend on one another.

As this book went to press, the United Nations and World Health Organization published its Joint Monitoring Project assessment on the state of water and sanitation connection in 2010. They were able to announce that the water Millennium Development Goal for 2015 had been reached, but that sanitation was still falling behind the targets.[19] They estimate that 780 million people lacked improved water and 2.5 billion improved sanitation and that these are forecast to fall respectively to 605 million and 2.4 billion by 2015. In contrast, the proportion with access to a household tap has fallen from 81% of the urban population in 1990 to 80% by 2010, with urban household access falling from 43% to 34% in Sub-Saharan Africa and 53% to 51% in south Asia. The report notes that 228 million people with household tap water rely on bottled water for domestic consumption. The number of urban dwellers relying on bottled water shot up from 26 million in 1990 to 192 million in 2010. In the report, people relying on bottled water are classified as being served by a household pipe, so stripping these numbers from the access data gives a slightly better picture regarding household access to trustworthy supplies. This means that urban household access (net of bottled water users) actually fell from 79% to 73% over the two decades.

It is time to set our sights higher. The United Nations is calling for a proposal seeking universal safe water and sewerage access by 2030.[20] The reality we face is that at current rates of progress, the OECD anticipates that in

2050, 240 million people will still depend on unsafe water and 1.8 billion will lack basic sanitation.[21] We stand at the cusp of optimism or pessimism. What could be as noble as ensuring that humanity has affordable and sustainable access to one of the most important elements of decent and civilised society? In an era of demographic and climatic change, the greatest change of all ought to be in our attitudes and ambitions.

---

[1] Biswas A K & Tortajada C,eds (2009) Impacts of Megaconferences on the Water Sector, Water Resources Development and Management, Springer-Verlag, Berlin, Germany

[2] Mount D T & Bielak A T (2011) Deep Words, Shallow Words: An Initial Analysis of Water Discourse in Four Decades of UN Declarations, UNU-INWEH, Hamilton, Ontario, Canada

[3] Biswas A K, Personal communication, 05–2011

[4] http://costsofwar.org/

[5] Lynas M (2011) *The God Species: How the Planet Can Survive the Age of Humans*, Fourth Estate, London, UK

[6] Gasson C (2012) The water sector and the Bengal famine of 1943, GWI Briefing, 89, 19th January 2012

[7] UNESCO IHE (2006) Urban water conflicts: An analysis of the origins and nature of water-related unrest and conflicts in the urban context. International Hydrological Programme / United Nations Educational, Scientific and Cultural Organization, Delft, the Netherlands

[8] World Bank (2004) Seawater and Brackish Water Desalination in the Middle East, North Africa and Central Asia: A Review of Key issues and Experiences in Six Countries, The World Bank, Washington DC, USA

[9] Loftus A (2011) Thinking rationally about water: review based on Linton's What is water?, *The Geographical Journal*, 177 (2) 186–188

[10] http://www.bluewaterbio.com/technology-hybacs-system.asp

[11] UNU-IWEH (2010) Sanitation as a Key to Global Health: Voices from the Field, UN University Institute for Water, Environment and Health, Hamilton, Canada

[12] OECD (2009) Managing Water for All: An OECD Perspective on Pricing and Financing, OECD, Paris, France

[13] Moyo D (2009) *Dead Aid: Why Aid Is Not Working and How There Is a Better Way For Africa*, Allen Lane, UK

[14] White S B (2011) Decentralising Water and Wastewater Systems: Lessons from the Telecom Industry, The Water Leader, 3, 51–52, June 2011, Institute of Water Policy, NUS, Singapore

[15] Gasson C (2011) The future is temporary, GWI Briefing, 26th May 2011

[16] Rogers P (2011) Water Governance: the Relevance of Price Policy, The Water Leader, 3, 34–35, June 2011, Institute of Water Policy, NUS, Singapore

[17] Biswas A K (2010) Water management in 2020 & beyond, Transforming the world of water: Global Water Summit 2010, Media Analytics, Oxford UK

[18] Danilenko A, Dickson E & Jacoben M (2010) Climate change and urban water utilities: challenges & opportunities. Water Working Notes No 24, The World Bank, Washington DC, USA

[19] UNICEF / WHO (2012) Progress on Drinking Water and Sanitation 2012 update, UNICEF & WHO JMP, Geneva, Switzerland

[20] Zafar Adeel, remarks at the World Water-Tech Investment summit, London, 28th February 2012

[21] Anthony Cox, remarks at the World Water-Tech Investment summit, London, 28th February 2012

Select Bibliography

Armstrong E L ed. (1976) *History of Public Works in the United States, 1776-1976*. American Public Works Association, Chicago, USA

Bakker K (2003) *An Uncooperative Commodity: Privatizing Water in England and Wales*. OUP, Oxford, UK

Bakker K (2010) *Privatizing Water: Governance Failure and the World's Urban Water Crisis*. Cornell University Press, New York, USA

Barlow M (2007) Blue Covenant: The Fight for Water as a Human Right. McClelland & Stewart, Toronto, Canada

Barlow M & Clarke T (2002). Blue Gold: The Fight to Stop the Corporate Theft of the World's Water. The New Press, NY, USA

Barty-King H (1992) *Water; The Book*. Quiller Press, London, UK

Bowen W R (2008) *Engineering Ethics: Outline of an Aspirational Approach*. Springer

British Medical Journal (2007) Medical Milestones Supplement. *Brit Med J* 344 Suppl 1.

Chartres C & Varma S (2010) *Out of Water: From Abundance to Scarcity and How to Solve the World's Water Problems*. *FT* Press, New Jersey, USA

Cosgrove W J & Rijberman F R (2000) *World Water Vision: Making Water Everybody's Business*. Earthscan, London, UK

Costanza R R et al (1997) The value of the world's ecosystem services and natural capital. Nature, 387(6230):255

Defra (2009) *UK Climate Change Projections 2009 – planning for our future climate*. Defra, London, UK

Falkenmark M & Lindh G (1976) *Water for a starving world*. Westview Press, Boulder, Co, USA

Falkenmark M et al (2007) On the Verge of a New Water Scarcity: A Call for Good Governance and Human Ingenuity. SIWI Policy Brief. SIWI, Stockholm, Sweden

Gassner, K Popov A & Pushak N (2009) Does Private Sector Participation Improve Performance in Electricity and Water Distribution? Trends and policy options No 6, The World Bank, Washington DC, USA

Gleick P (2010) *Bottled and Sold: The Story Behind Our Obsession With Bottled Water*, Island Press, Washington DC, USA

Gleick P H & Palaniappan M (2010) Peak water: Conceptual and practical limits to freshwater withdrawal and use. Proceedings of the National Academy of Sciences, published online before print May 24, 2010; doi: 10.1073/pnas.1004812107

Gulyani S, Talukdar D & Kariuki R M (2005) Water for the Urban Poor: Water Markets, Household Demand, and Service Preferences in Kenya. Water Supply and Sanitation Sector board Discussion Papers Series, No 5, The World bank, Washington DC, USA

GWM 2011 (2010) Global Water Markets 2011. Media Analytics, Oxford, UK

Hall D & Lobina E (2008) Sewerage works: Public investment in sewers saves lives. PSIRU, London, UK

Halliday S (1999) *The Great Stink of London. Sir Joseph Bazalgette and the Cleansing of the Victorian Metropolis*. Phoenix Mill: Sutton Publishing.

HDR (2006) *Human Development Report 2006. Beyond scarcity: Power, poverty and the global water crisis.* UNDP, New York, USA

Howard G & Bartram J (2003) *Domestic Water Quantity, Service Level and Health.* WHO, Geneva, Switzerland

Hutton G & Haller L (2004) *Evaluation of the costs and benefits of water and sanitation improvements at the global level. World* Health Organization, Geneva, Switzerland Investing

IFC (2009) Safe Water for All: Harnessing the Private Sector to Reach the Underserved. IFC, Washington DC, USA

IPCC (2008) *Climate change and water: IPCC Technical Paper IV.* IPCC, Geneva, Switzerland

Kauffmann C (2009) *Private Participation in Water Infrastructure: OECD Checklist for Public Action.* OECD, Paris, France

Lomborg B (2001) *The sceptical environmentalist: Measuring the Real State of the World.* Cambridge University Press, Cambridge, UK

Lynas M (2011) *The God Species: How the Planet Can Survive the Age of Humans.* Fourth Estate, London, UK

Marin (2009) Public-Private Partnerships for Urban Water Utilities: A Review of Experiences in Developing Countries World Bank, Washington DC, USA

Matti Kummu et al (2010) Is physical water scarcity a new phenomenon? Global assessment of water shortage over the last two millennia. *Environ. Res. Lett.* 5 034006

Meybeck M (2003) Global analysis of river systems: from Earth system controls to Anthropocene syndromes. Philosophical Transactions of the Royal Society, B, 2003, 358, 1935-1955

Mount D T & Bielak A T (2011) *Deep Words, Shallow Words: An Initial Analysis of Water Discourse in Four Decades of UN Declarations*. UNU-INWEH, Hamilton, Ontario, Canada

OECD (2007) *Financing Water Supply and Sanitation in ECCA Countries and Progress in Achieving the Water-Related MDGs*. OECD, Paris, France

OECD (2009) *Managing water for all: An OECD perspective on pricing and financing*. OECD, Paris, France

OECD (2009) *Report on measures to cope with over-fragmentation in the water supply and sanitation sector*. Report by Kommunalkredit Public Consulting, OECD, Paris, France

OECD (2010) *Pricing Water Resources and Water and Sanitation Services*. OECD, Paris, France

Payen G (2011) Worldwide needs for safe drinking water are underestimated. Aquafed, Brussels, Belgium

Pearce F (2006) *When the Rivers Run Dry.* Eden Project Books, London, UK

Prüss-Üstün A, Bos R, Gore F & Bartram J (2008) *Safer water, better health: costs, benefits and sustainability of interventions to protect and promote health.* World Health Organization, Geneva, Switzerland

Shiklomanov I H & Rodda J (2003) *World Water resources at the Beginning of the 21st Century.* Cambridge University Press, Cambridge, UK

Slattery K (2003) What Went Wrong? Lessons from Cochabamba, Manila, Buenos Aires, and Atlanta. Water & Wastewater Annual Privatization Report, The Reason Foundation. Los Angeles USA

Solomon S (2010) *Water: The Epic Struggle for Wealth, Power, and Civilization.* Harper Collins, NY, USA

Stern N (2007) *The Economics of Climate Change: The Stern Review,* Cabinet Office – HM Treasury, London, UK

Transparency International (2008) *Global Corruption Report 2008: Corruption in the Water Sector.* Cambridge University Press, Cambridge

Tremolet S (2010) *Innovative Financing Mechanisms for the Water Sector.* OECD, Paris, France

Tremolet S (2011) *Benefits of Investing in Water and Sanitation.* OECD, Paris, France

UNCHR (2010) The Right to Water. Office of the United Nations High Commissioner for Human Rights, Geneva, Switzerland

UNESCO IHE (2006) *Urban water conflicts: An analysis of the origins and nature of water-related unrest and conflicts in the urban context*. International Hydrological Programme / United Nations Educational, Scientific and Cultural Organization, Delft, the Netherlands

UN-Habitat (2003) *The Challenge of Slums: Global report on human settlements, 2003*. UN-Habitat / Earthscan, London, UK

UN-Habitat (2007) *Enhancing urban safety and security: Global report on human settlements 2007*. Earthscan, London, UK

Veolia Water (2011) Finding the Blue Path for a Sustainable Economy. A White paper by Veolia Water. Veolia Water, Chicago, USA

Vörösmarty C J et al (2010) Global threats to human water security and river biodiversity. Nature 467, 555-61

WaterAid (2006) Total sanitation in South Asia: the challenges ahead. WaterAid, London, UK

WHO (2003) *Emerging Issues in Water and Infectious Disease*. World Health Organization, Geneva, Switzerland

WHO (2005) Making water part of economic development: The economic benefits of improved water management and services. Stockholm International Water Institute and World Health Organization, Stockholm, Sweden

WHO / UNICEF JMP (2010) *Progress on Drinking-water and Sanitation 2010 update*. WHO, Geneva, Switzerland

WHO (2011) *Guidelines for Drinking-water Quality*, Fourth Edition. WHO, Geneva, Switzerland

Winpenny J (2003) *Financing Water For All: Report of the World Panel on Financing Water Infrastructure*. World Water Council

Winpenny J et al (2010) *The wealth of waste: The economics of wastewater use in agriculture*. FAO water reports 35, FAO Rome, Italy

World Bank (2002) Water Tariffs and Subsidies in South Asia. World Bank, Washington DC, USA

WWDR 2 (2006) *World Water Development Report 2: Water: A Shared Responsibility*. UNESCO, Paris, France

Acknowledgements

This book came about from friendships forged in west Wales. The idea about needing to tell the story of urban water came when David and Clare Hieatt invited me to be one of the Do Lecturers for 2010. Glen Peters and Brenda Squires held a writers' retreat at Rhos-y-Gilwen later that year where stories were developed. Richard Davies and Gillian Griffiths at Parthian Books told me to set aside 'the novel that time forgot' and take this on. Kathryn Gray made a host of valuable points while editing the manuscript, with diplomatic panache.

Many ideas and observations have evolved from a plethora of meetings, workmanlike and convivial, with water professionals, consultants, academics and writers. In particular, Sophie Tremolet, David Johnstone, Mark Lane, Karen Bakker, Philippe Rhoner, Hans Peter Portner, Arnaud Bisschop, James Hotchkies, Michael Deane, Jean Pilloud, Gerard Bonnis, Urooj Amjad, Geraldine Dalton,

Angela Whelan, Asit Biswas, David Suratgar, Jim Winpenny, Gareth Roberts, Seungho Lee, Tony Allan, Jack Moss, Ian Elkins and Christopher Gasson have shared many thoughts over the years. I have also gained valuable insights through consulting projects with a variety of clients and I am grateful for the support they have given me.

The title *The Sound of Thirst* comes from a line in Gillian Clarke's *A Recipe for Water* (Carcanet, 2009) and her long association with Glas Cymru Cyf has enlightened Wales' water.

Above all are my three soundboards: Polly, Bethan and Trystan. Polly is my now, my loving Devil's Advocate. Bethan and Trystan are our future.

Index